ELECTRO-PLATING
FOR THE
AMATEUR

ELECTRO-PLATING

FOR THE

AMATEUR

L. WARBURTON

ARGUS BOOKS
LIMITED

Argus Books Ltd
P.O. Box 35
Hemel Hempstead,
Herts, England

First published 1950

Second Edition 1955

Third Edition 1963

Second Impression 1969

Third Impression 1970

Fourth Impression 1972

Fifth Impression 1973

Sixth Impression 1975

Seventh Impression 1976

Eighth Impression 1977

Ninth Impression 1980

Tenth Impression 1981

Eleventh Impression 1982

Twelfth Impression 1984

ISBN 0 85344 056 5

Printed in Great Britain by A. Wheaton & Co. Ltd., Exeter

CONTENTS

PREFACE

I N presenting this book, I have only two objects in mind. Firstly, to try and provide some really useful information about small-scale electro-plating and secondly, whilst avoiding technicalities, to give emphasis to those essential details which must be carefully attended to if successful work is to be carried out.

The methods described are not just the result of adapting commercial processes to home-workshop conditions, but are rather the reverse, in that they represent the outcome of several years of experiment, starting more or less from scratch, during which time a reliable method has been evolved in respect of each process described.

The subject of chromium plating has purposely been left until near the end of this book, and I hope that those of my readers who are novices will not attempt chrome until the processes in copper and nickel have been mastered. The final chapter is devoted to a description of two simple methods of "anodising," which is a comparatively recent development in the preservation and decoration of aluminium and which, in spite of its simplicity, is highly satisfactory; more so, in my opinion, than the actual electro-plating of aluminium and its alloys.

Regarding the various pieces of equipment which I have described, I do not suggest that they are ideal, but I do feel that they have been reduced to their simplest form without serious loss of efficiency. I sincerely hope that, having had the patience to equip himself with a small plant of the type described, the reader will derive as much pleasure and profit from this fascinating pursuit as I myself have enjoyed over the last few years.

L.W.

1   INTRODUCTION AND GENERAL CONSIDERATIONS

THE art, or science, of electro-plating has reached a very high degree of perfection in present-day industry and has innumerable commercial applications, quite apart from merely decorative or protective purposes, and indeed the electro-deposition of various metals in some branches of industry, notably in printing, has been developed into very specialised channels.

This book is not intended to deal with the specially applied branches of the subject, and the various types of process to be described are set forth only with a view to providing a decorative and a protective coating of metal. In addition, notes on electro-forming are given at the end of Chapter 7. The following chapters are intended to provide the amateur engineer with what is hoped will be sufficient data, not only to carry out successful electro-plating in the small workshop, but also to provide himself with the essential tools of the trade, i.e., the electrical equipment and plating tanks.

When one wishes to set up a small workshop to carry out model engineering, one can quite easily go forth and purchase a complete outfit of tools and equipment and get on with the job. Not so, however, in the case of the prospective electro-plater, because so far as the author is aware the only plating plant obtainable is on a far bigger scale than anything required by even the most enthusiastic amateur. Consequently it has been thought fit to devote a very considerable portion of this book to a detailed discussion of a suitable size of plant, together with details as to how such a plant can be assembled in the small workshop.

For obvious reasons it has been assumed that the reader is one of the band of very practical people known collectively as model engineers, and bearing this in mind, it has also been assumed that he is quite able to look after the purely mechanical processes involved in setting up a suitable plant. Those who are not in that position, and are yet interested in the present subject can no doubt obtain the necessary assistance in the mechanical details.

In setting out to equip oneself with a small plating plant, the primary consideration will be the size of job to be undertaken because upon this depend all the other matters to be taken into account. In dealing with electro-deposition, the " size " of the job means, first and last, the area of metal in contact with the liquid, and on which a plate is to be formed. For example, a circular rod $\frac{1}{16}$ in. in diameter and 13 in. long, has roughly the same total surface area as a penny and is, electrically speaking, the same size. It is obvious however that it will need a very much larger tank to accommodate it. It will be seen, therefore, that for the purposes of this book some limit must be set both to surface area and to longest dimension of the work to be plated, and for the sake of simplicity it has been assumed that the following will represent the maximum measurements of any work undertaken :

 a Longest dimension—10 in.

 b Surface area not to exceed—160 sq. in.

These measurements correspond with, say, a 10 in. disc, both sides being treated at once, or with a 5 in. cube.
The tanks described, as also the electrical equipment, are designed with these limits in mind, and if larger objects require to be plated, the necessary all-round increase in size of equipment will have to be made.

2

BASE METALS

The terms "base metal," or simply "base" will be used frequently throughout the text, and it should be explained that these terms refer to the actual article to be plated, which shall be our next consideration.

Different bases require widely differing treatments to prepare them for receiving a deposit, and there is no doubt that the initial preparation of the base is by far the most important process in the series of operations. The author has never experienced so strong a feeling of disappointment as when an otherwise perfect plate has blistered and peeled off during the final polishing. For this reason, the reader's attention is invited most particularly to Chapter 4, which deals with the initial preparation of the base, and meticulous attention to detail at this stage cannot be too strongly recommended.

The following types of bases are dealt with in this book, and will probably cover most of the field, so far as the amateur is concerned.

1. Copper.
2. Cast iron.
3. Wrought iron and steel.
4. Brass and tin-bronze.
5. Lead and pewter (tin-lead).
6. Lead-tin-zinc-antimony alloys (called Britannia metal).
7. Zinc and galvanised ware.
8. Tinplate (on iron).
9. Aluminium.
10. Non-conductors such as glass and plastics.

In view of the admitted difficulties experienced, even by experts, in satisfactorily plating the light alloys, they have been completely left out of this book. The details given for plating pure sheet aluminium are practically useless when applied to most of the alloys of this metal.

SOURCE OF SUPPLY OF CURRENT

With the exception of the next few paragraphs, and a brief reference to generation in a later chapter, it has been assumed throughout this book that the reader has a supply of current laid on from the public main, and the conversion of such current, both D.C. and A.C., is discussed at some length in Chapter 2.

The use of batteries is expensive and often somewhat disappointing, but against this it can be said that the author has accomplished some excellent silver-plating on a set of tea-spoons with the aid of nothing more than a 6 volt motor-cycle battery, an ammeter and a length of iron wire for a resistance. If there is no alternative but to use batteries, it is recommended that a heavy duty starter-battery of at least 45 ampere-hours be employed, and in this case all that will be required by way of additional electrical equipment is an ammeter in the negative lead, and a variable resistance in the positive, with some means of tapping off a less voltage if required. The resistance described in the next chapter is quite suitable for use with a battery.

Before leaving the subject of batteries as a source of current, a point well worth noting is that economy of current can be effected by plating at a somewhat lower rate than those given in the tables. In all electrolytic processes there tends to be some decomposition of the water in which the various salts are dissolved. This proceeds at a higher rate with higher current densities, and is a total loss of current so far as deposition of metal goes. It follows that using lower current densities will ensure that a greater percentage of the total current used will be usefully employed in depositing metal on the cathode, but of course a proportionately longer time will be required to deposit the same weight of plate. It is therefore suggested that battery currents be kept down to two-thirds of the values laid down in the tables, the time of plating being increased by one-half.

4

It must be emphasised that none of the foregoing remarks can apply in the case of chromium plating. In this the current density is very critical and must be as stated; any considerable departure from the given values will ruin the plate. In any case, owing to the very high currents required for chroming, only the smallest bases can successfully be treated by battery power.

TYPES OF PLATE TO BE CONSIDERED

The scope of this little book is necessarily limited, and it has been thought advisable to give detailed information about a limited number of the commonest and most useful plates, rather than a sketchy outline of a great many.

Plating in the following metals only will, therefore, be discussed :

a Copper.

b White brass (known also as Hard brass).

c Zinc.

d Nickel.

e Silver.

f Cupro-nickel ("German silver").

g Chromium.

h Gold.

Such plates as cadmium, indium, rhodium and platinum, etc., are ignored, as being quite outside the usual requirements of the amateur.

DETERMINATION OF SURFACE AREA

The amount of current which is to be passed through any electrolyte for the purpose of depositing metal is determined by two main factors. Firstly by the nature of the solution containing the metal, and secondly by the area of surface in contact with the solution, and which is to receive the plate.

It must be admitted at the outset that the accurate determination of surface area can, in some instances, prove very difficult indeed, and in the case of very intricate surfaces may be frankly impossible. Fortunately, with the exception of chrome, the matter is not very critical, and errors of from 20 to 40 per cent. are usually well tolerated. The appearance of the work during actual plating is quite a good indication of whether the current density is approximately correct, and notes on this are given in the details of electrolytes in Chapter 5.

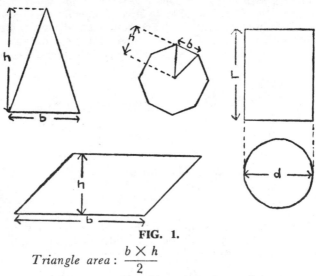

FIG. 1.

Triangle area : $\dfrac{b \times h}{2}$

Polygon area : $\dfrac{b \times h \times number\ of\ sides}{2}$

Cylinder (hollow) outside area : $\pi \times d \times L$

Cylinder (solid) area : $\pi \times \left\{ (d \times L) + \frac{d^2}{2} \right\}$

Parallelogram area : $b \times h$

For convenience, the calculation of surface area of a few simple geometrical figures is given in Fig. 1, and it is suggested that for irregular shapes, a fairly close estimate can usually be made. An instance of this is given in Fig. 2, where a pewter sugarbowl is "averaged out." Similar treatment will often suffice for objects of different shape.

It is assumed that the reader can calculate the surface area of cubes, rectangular blocks, etc., bearing in mind that the required figure is the sum of all the surfaces. In

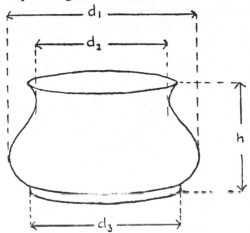

FIG. 2.

Approximate area is obtained by treating the sugar-bowl as a cylinder, whose dimensions are :-

$$D \ (diam. \ of \ cylinder): \frac{d_1 + d_2}{2}$$

and area is therefore : $(\pi \times D \times h) + \left(\pi \times \dfrac{d_3{}^2}{4}\right)$

This gives external area only. Internal area is roughly equal for thin bases.

this connection, it should be borne in mind that in the case of hollow articles, where the plating solution has access to

7

the inside, the area of the inner surface must also be taken into account. Where the material of the base is thin, the inside area will be nearly equal to the outside.

In cases where blocking-out is done, the area blocked-out must be deducted from the total area, as the solution has no electrical contact with the metal at this point. The uses and methods of such blocking-out processes are discussed in Chapter 4.

There remains one final point to be made regarding the calculation of surface area. It often happens that one side only of a base is required to be plated, for example, a head-lamp reflector. Rather than go to the trouble of blocking-out the blind side, quite a useful plan is to employ only one anode, this facing the working side of the base. Under these circumstances, a certain amount of metal will be deposited on the blind side, but very much less than on the working side, and, although the correct calculation of what current to allow depends on innumerable factors, the author has found by experience that a happy medium can be found, in most cases, by reckoning one full area of the working side, plus half the area of the blind side. The only time when this system gives trouble is in chromium-plating, when two anodes at least must be used, and the full area taken into account.

If the reflector referred to above is being silvered, the time spent in thoroughly blocking-out the blind side will be well repaid by the saving in silver. It is an expensive metal.

2 *THE SUPPLY OF CURRENT*

METALS can be electro-deposited only by a direct current. Alternating current will not, under any circumstances, produce an electrical plate, hence it is necessary to arrange for a supply of direct current of small voltage and relatively large amperage.

The public supply mains in this country are almost entirely alternating, and it is with this type of supply that the present chapter is chiefly concerned. A brief note regarding D.C. mains is included at the end of the chapter.

In the case of A.C. mains, the supply voltage is immaterial, as a transformer is invariably used to convert the current to the correct voltage. The only effect which the supply voltage will have is to determine the ratio between the number of turns in the primary and secondary windings. It being outside the scope of this book to give detail of windings, etc., it will perhaps suffice to state that a transformer will be required with primary to suit the mains, and with secondary output as follows :

    a One main continuous winding, tapped at 4, 6, 8, 10 and 12 volts, capable of giving a sustained current of 10 amperes for several hours.

    b If required, a subsidiary separate winding of 4 volts at $\frac{1}{2}$ amp. to light a pilot bulb.

    c A third winding of correct voltage and current to supply the small motor to be used for agitating the contents of the plating tank (see Chapter 3).

At the time of writing there is an ample supply of ex-Government transformers at a reasonable price, which meet the requirements nicely, and when purchasing, it is a good fault to err on over-size rather than under-size.

The transformer secondary switch is similar in construction to the resistance switch described later in this chapter, except that only five contact studs are required. A blind stud must, however, be placed between each live one, to prevent the arm shorting the transformer secondary tappings by making contact with two live studs at once. Connections are shown in Fig. 3. Under no circumstances

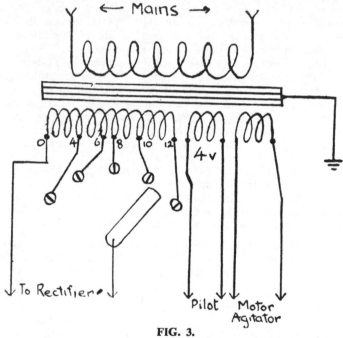

FIG. 3.

The tappings for motor, as shown, should be of a voltage to suit the motor fitted.

should an "auto-wound" transformer be used, as this would render the whole plating plant "alive."

ROTARY TRANSFORMERS

These consist of an electric motor, run direct from the mains, coupled to a low-voltage direct current generator.

Whilst such a converter, as they are called, can be made to run on either D.C. or A.C. mains, its use is confined almost entirely to D.C., as A.C. can be stepped down and rectified much more cheaply and conveniently by means of a static transformer and rectifier.

Accordingly, the rotary transformer is dealt with at the end of this chapter, under D.C. mains section.

RECTIFICATION OF CURRENT

The next most important item is the rectifier, and for the small plant under consideration, either a copper or a selenium rectifier will be suitable. The one chosen should be capable of rectifying 12 volts, and of passing 12 amps. without undue heating, and should for preference be of the double type for bridge connection. This gives full-wave rectification of the current, and is technically advantageous. For plating purposes, no smoothing apparatus is required, except in the case of chromium; *see chapter* 6.

If a bridge type rectifier is not obtainable, the alternatives are either to use a single rectifier for half-wave rectification, or better, two such rectifiers for full wave. Connections for these various arrangements are detailed in Fig. 4.

THE VARIABLE RESISTANCE

The next item of electrical equipment to be considered is the variable resistance, which controls the current passing through the plating tank.

This is a very important instrument, and if not carefully selected, or exceptionally well made, can lead to endless trouble, either through over-heating or through faulty contacts. The author has tried many different types and

11

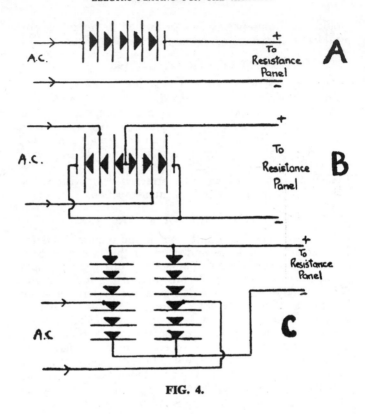

FIG. 4.

patterns, and has eventually come to the conclusion that the " step " type, with radial switch, is the most satisfactory. It is also fairly easy to make, and its only disadvantage lies in the fact that currents cannot be controlled to a very fine limit. For instance, it may happen that the required current is calculated to be, say, 6.5 amps. On trying various positions of the radial switch, quite possibly the operator is forced to choose between, say, 6 and 6.8 amps.

12

As previously mentioned, however, the current values are not critical, and the operator would choose the nearer value, i.e. 6.8 amps. On the assumption that the reader has some skill in simple engineering, the following notes on construction of a suitable resistance should be quite adequate to ensure that a satisfactory instrument is produced.

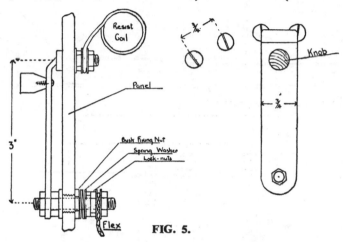

FIG. 5.

The general lay-out for single-panel control is shown in Fig. 6 and details of the resistance panel separately in Fig. 5. The dimensions given, although by no means rigid, are quite satisfactory, and they should not be reduced by any large amount, or difficulty may be experienced in accommodating the coils of the resistance.

The portion of panel carrying the resistance switch is either of slate or metal. In the latter case, the rotor spindle bearing and all contact-studs must be thoroughly insulated with mica or porcelain. The remainder of the panel can be of varnished wood.

The ideal resistance has a very large number of studs, but in practice, ten will be found to give adequate control,

13

REAR.

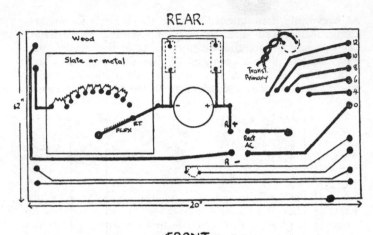

FRONT.

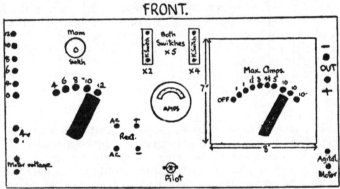

FIG. 6.

Combined panel, with all the controls.

especially as further control is obtainable by selections of various transformer secondary voltages. The construction of the switch panel is quite straightforward, and should present no difficulties.

14

The studs are brass cheese-headed machine screws, either $\frac{1}{4}$ in. Whit. or equivalent diameter in finer thread. They are securely fixed in their respective holes in the panel by a washer and nut at the back, and each carries two brass washers and a further nut for securing the resistance wires. The slots in the heads should be arranged radially, i.e., to point towards the rotor spindle. This latter is either copper or brass strip of $\frac{3}{4}$ in. width and about $\frac{1}{16}$ in. thickness, and is carried on a brass spindle passing through a brass or steel bush, which is in turn secured to the panel. This latter may be fixed in position either by means of a small flange with external thread and nut, or by a wide flange bolted to the panel.

The rotor spindle is fitted carefully through this bush, so that when the fixing nut and lock-nut are screwed up, there is a good pressure of the arm on the studs, which pressure is maintained by the springiness of the arm itself. This arm, by the way, should be of sufficient width to rest on two studs at once. Contact to the rotor through the bush is not reliable when dealing with large currents, and a flexible connector is recommended. This consists of a a length of heavy-duty mains flex, and is secured to the rear end of the spindle by a further pair of brass washers and a nut. The free end of the flex is then fixed to a terminal marked RT in Fig. 6, leaving an adequate amount of slack to be taken up on rotation, but at the same time making sure that under no circumstances can the flex come in contact with the resistance coils.

Fig. 7 is a diagram of the values of the resistances to be connected between each stud, and Table 1 shows the various resistance wires available, and their characteristics. From this table can be calculated the length of any particular wire for any required resistance.

A word here on the improvisation of resistances from a mains heater element will perhaps be useful. Heater elements of 240 volts and 1,000 watts rating are quite freely obtainable, and can be adapted to make the various sections as shown in Fig. 7. Such an element will pass

15

just over 4 amps. at a bright red heat, but this temperature is of course not permissible for the apparatus under
discussion. If the current flowing through the wire is
restricted to 1½ amps. the temperature rise will not be too

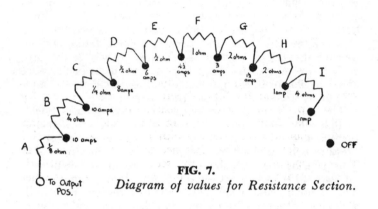

FIG. 7.
Diagram of values for Resistance Section.

great, and a single strand of the wire can therefore be used
for the sections marked G, H and I in Fig. 7. Two strands
in parallel will suffice for section F, and three strands in
parallel for section E. It is quite easy to calculate the
length of wire required for sections G, H and I, bearing in
mind that the entire element, as purchased, has a total
resistance of 55.2 ohms. Hence it will be seen that
section I, which is to be 4 ohms, will need to be

$$\frac{4}{55.2} \text{ which is roughly } \frac{1}{14} \text{ total length.}$$

Sections G and H will have to be $\frac{1}{28}$ of the total length.

16

Section F, although only half the resistance of G, is composed of two wires in parallel, and each of these, therefore, will have to be as long as G, i.e., 1/28 of total length. Similarly, section E is to be $\frac{1}{2}$ ohm, but in this case three strands are used, each of which must be $1\frac{1}{2}$ ohms, or 1/37 of the total length.

The remaining sections, A, B, C and D, cannot be built up conveniently from a mains heater element, owing to the high current-carrying capacity required, and the author has found that iron wire such as is used for light fencing purposes is quite suitable for these sections. This wire is almost invariably galvanised, and the zinc should be removed by careful filing, and sand-papering, for if allowed to remain, it will appreciably alter the resistance.

The final diameter of the wire after removing the zinc coating should be ascertained with a micrometer, and the resistance determined by reference to Table 1. From this the appropriate length for each section can be calculated. The resistance of this wire being very low, these sections will be quite bulky, and the reader should not make any attempt to economise in space by choosing a wire which, whilst thinner, would be unable to carry the necessary current. The value of current which each size of wire can comfortably carry is shown in the table, and must not be exceeded or otherwise over-heating will occur.

Having selected and cut to size all the sections, they will have to be prepared for fitting to the studs as follows.

The ends of all the wires are thoroughly cleaned by scraping, and each section is then formed up to the shape shown in Fig. 8, with the exception of sections A, B, C and D, which will require longer " legs " to carry them further back from the panel, on account of their larger bulk. To form, the wires are wound around a suitable mandrel, the actual diameter of which will of course be determined by the length of the respective wires, and may not be the same for each. The object is to reduce the wires to a convenient shape for fitting, and details of this are best left to the ingenuity of the reader. There are only two snags to

17

watch out for : firstly, no two adjacent turns of any coil
must be allowed to come in contact and, secondly, no two
coils must touch each other, except at the point where they
are joined by the stud nuts.

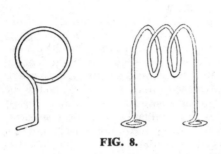

FIG. 8.

*The number of turns in the coils will
depend upon the length of wire in each.*

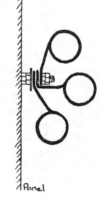

FIG. 9.

*Disposition of triple-coils for Section
E. Double coils for Section F are arranged similarly.*

The loops at the end of each section must be made a good
round fit on the stud shanks, and in the case of sections E
and F, which have more than one coil each, particular
attention must be paid to the danger of squeezing out one
or more loops when tightening the clamping nut. The
twin and triple coils of these two sections should be formed
separately, and spread out fan-wise as shown in semi-
diagrammatic form in Fig. 9.

This is a much better plan than twisting the wires
together before forming the coil, as any heat produced is
much more readily dissipated.

As will be apparent from Fig. 7, the beginning of coil A
does not go to a stud, but directly to the terminal marked

OUT in Fig. 6, thus ensuring that at least $\frac{1}{8}$ ohm is always in circuit.

A useful tip, which helps to guard against overloading of any sections of the resistance, is to mark clearly over each stud the maximum current which that particular section is intended to pass. These values are shown in Fig. 7, and should not be exceeded. In other words, when the rotor arm is in contact with any particular stud, the whole current in the circuit must not be allowed to exceed the figure which is marked above that stud. If this should appear to be necessary at any time during actual use, it is an indication that too high a voltage is being used, and the transformer secondary tapping should be reduced accordingly. In practice, it will rarely be necessary to use more than 6 volts, except when chroming, but the actual voltage required will depend largely upon the internal resistance of the rectifier.

To return to the resistance, all that is now required is to fit a suitable knob or handle to the rotor arm, and two stops, one at each end of the arc of studs, to prevent the arm from slipping off. The tenth stud, it will be seen, is a blind one, and serves to switch off the plating current, whilst making any adjustment in the tank. The arm should always be switched over to this " off " position before switching the mains current on or off.

METERS

The only meter actually necessary is an ammeter, although the addition of a voltmeter is considered by some workers to be an advantage. Personally, although the author has a voltmeter included in his plant, it must be many months since any notice was taken of it. After all, it is the current passing in the circuit which matters, and, provided all connections are good, the voltage across the electrodes must be proportional to the current in any given solution, under normal working conditions.

19

TABLE I   Resistances of various wires suitable for the present case

Wire	S.W.G.	Diameter in inches	Feet per Ohm	Ohms per Foot	Max. Current for Present Case
IRON ...	15	0·072	82	0·012	17 amps.
	16	0·064	67	0·015	12 ,,
	17	0·056	49	0·021	7 ,,
	18	0·048	37	0·028	4 ,,
	19	0·040	28	0·036	3 ,,
NICKEL ...	15	0·072	55	0·018	10 amps.
	16	0·064	44	0·022	8 ,,
	17	0·056	34	0·029	6 ,,
	18	0·048	28	0·036	5 ,,
	19¾	0·040	22	0·045	4 ,,
	20	0·032	18	0·056	3 ,,
20% NI NICKEL-SILVER	16	0·064	2	0·084	8 amps.
	17	0·056	9	0·107	6 ,,
	18	0·048	7½	0·136	4½ ,,
	19	0·040	6	0·168	3 ,,
	20	0·032	4¾	0·222	2½ ,,
NI-CHROME	15	0·072	5	0·208	8 amps.
	16	0·064	4	0·259	6¼ ,,
	17	0·056	3	0·333	5 ,,
	18	0·048	2½	0·421	4 ,,
	19	0·040	2	0·520	3 ,,
	20	0·032	1½	0·659	2½ ,,

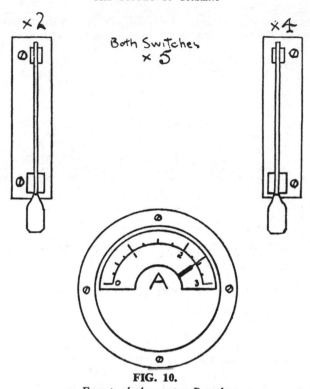

FIG. 10.
Front of Ammeter Panel.

The ammeter itself is required to give accurate readings over a long scale, say, from ½ amp. to 12 amps., and this is unusual in a single dial meter. The solution to this is quite simple, and consists in employing an ammeter reading from 0 to 3 amps., combined with suitable shunt wires, and the setting up of such an instrument could be tackled as follows (Figs. 10 and 11).

The ammeter is mounted on the instrument panel in any convenient way, and a thick copper wire, of about 14 s.w.g.

is firmly connected to one terminal, and run to the terminal marked R.pos. on the Panel (Fig. 6), where it is securely fixed by a really sound job of soft-soldering. This is repeated at the other terminal of the meter, running the wire, in this case, to the terminal marked RT on the

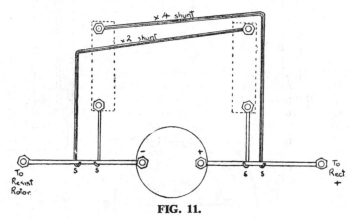

FIG. 11.

Rear of Ammeter Panel. Joints marked " S " are soldered.

resistance switch (Figs. 6 and 11). On each side of the meter is mounted a knife switch, of the common pattern, with self-cleaning contacts. These switches must have all metal parts made of copper, and be of ample size; they cannot, theoretically, be too large, although for practical purposes, those with a blade $\frac{1}{2}$ in. wide and $\frac{1}{16}$ in. thick would serve. The contact must be really close and positive.

The nearest terminal of each switch is then firmly connected to the respective conductors on each side of the meter, again with 14 s.w.g. copper wire, and again soft-soldered thoroughly. Over the left-hand switch, the panel is then marked " X2," and over the right-hand one, " X4." In the centre, over the meter, the panel is marked " Both Switches X5."

There remains now, only the calibration of the meter, by the fitting of two shunt wires, and this is done in the following way. The transformer, rectifier, resistance and meter are all connected up, as in Fig. 6, and the variable resistance set at maximum resistance. The transformer output is set at 4 volts, and both knife switches are closed. The output terminals on the panel, marked " + and -," are then shorted by a piece of 14 s.w.g. copper wire, and the resistance is adjusted to give a 3 amp. deflection on the meter. If this cannot be done accurately at 4 volts, the 6 volt tapping should be tried, adjusting the resistance again. When a 3 amp. current is obtained, switch off the mains, leaving the controls as they are. To the free terminal of the left-hand switch (" X2 "), a length of 22 or 24 s.w.g. copper wire is secured, and the other end is pressed tightly against the conductor connected from the terminal marked R.pos. to the meter. This contact should be made as perfect as possible, by squeezing the wires together with pliers. By trial and error, the correct length of wire is found which, when current is switched on, will give a reading of exactly $1\frac{1}{2}$ amp. on the meter. Without cutting the excess off the shunt wire, it is soft-soldered to the conductor, and the reading again checked. It may be a little low, owing to the improvement in the connection when soldered, and if so, it is unsoldered, and a little more length allowed. When the correct length is found, the shunt is coiled into a spiral and finally soldered into place, and the knife-switch opened. If, during this fitting, it has been found that the required shunt is un-reasonably long, it should be replaced by a thinner wire, and vice versa. To the free terminal of the right-hand knife switch, another shunt wire, this time of 18 or 20 s.w.g., is fixed, and the whole of the above process repeated, this time, of course, running the free end of the shunt wire to the conductor joining the meter to the " RT " terminal on the resistance panel, so that the reading on the meter is reduced to $\frac{3}{4}$ amp. when the right-hand switch is closed, and the other one opened.

23

The panel is now complete, and should be tested once more as follows.

With the " X2 " switch closed, and the other open, the current is adjusted to read 3 amps. The " X4 " switch is closed, and the current should read $1\frac{1}{6}$ amps. Leaving the " X4 " switch closed, the " X2 " switch is opened, and the reading should then increase to $1\frac{1}{2}$ amps.

Operation is quite simple; for currents up to 3 amps. both switches are open. For currents between 3 and 6 amps., the " X2 " switch is closed; between 6 and 12 amps., the " X2 " switch is open and the " X4 " closed, and for currents between 12 and 15 amps., both switches are closed. To find the actual current, the meter reading is multiplied by either 2, 4 or 5, according to the position of the two switches.

D.C. SUPPLY MAINS

If supply mains of, say, 12 to 20 volts D.C. were available, all problems connected with voltage reduction and rectification would disappear. Unfortunately this is never the case, and mains supply at less than 100 volts is indeed rare. By far the most common voltages are those around the 200 to 250 mark, and this fact introduces problems.

It is quite possible, from a theoretical standpoint at least, to employ D.C. current direct from a 250 volt main for electro-plating, simply by inserting a suitable resistance. The matter, from a practical standpoint, however, is not so simple. Firstly, under those circumstances, all the apparatus, including the plating tank, would be " alive," and under most conditions of supply it would be dangerous to handle any part of the apparatus. Insulation, too, would be extremely difficult to arrange, particularly so in view of the presence of liquid, and moist fumes. Perhaps the greatest drawback, however, to direct use of D.C. mains in this way, is the cost.

24

Let us consider for a moment the case of a large article being plated, say one requiring 15 amps., at a pressure of 6 volts. The actual current consumed in the plating tank would therefore be 90 watts, and even allowing for losses the total energy drawn from an A.C. main, using a static transformer and a rectifier, would amount to only about 120 to 130 watts; quite a modest figure.

Now let us consider the same job being done by direct connection (via suitable resistance) to a D.C. main at 250 volts. A current of 15 amps. would flow through the whole circuit, and consequently through the supply meter, and this latter would therefore register a total energy consumption of 15×250, or 3,750 watts, nearly thirty times as much current to do the same job. In addition to this, the initial cost of a resistance capable of absorbing 244 volts at a current of 15 amps. would be enormous, to say nothing of the size, which would likewise be enormous. Thus it is quite reasonable to dismiss this method as impracticable, and there remains the alternative of using a rotary transformer, or convertor, as it is commonly called.

As previously mentioned, this machine· consists of a motor coupled to a dynamo of suitable output, and many such convertors are on the market at present. Initial cost is rather high, but upkeep costs are remarkably low, and a machine of reliable manufacture is quite trouble-free, the only maintenance required being an occasional drop of oil and very infrequent renewal of the brushes. The reader who has the necessary skill could make an excellent converter by coupling a suitable electric motor to a car dynamo. These dynamos, by the way, lend themselves very well indeed to plating work, as the third brush regulation enables output voltage to be controlled. They do, however, necessitate an accumulator to be connected always across the output in order to stabilise the voltage and to prevent excessive " building up," with consequent damage to the field coils.

An arrangement which the author has seen working, and which works exceptionally well, consists of a $\frac{1}{8}$ h.p.

25

motor (shunt wound for fairly constant speed) driving a 12 volt car dynamo. This latter had its third brush fixed at maximum position, and the lead from it to the field coil was broken, and a variable resistance of 0 to 30 ohms inserted. The dynamo was connected to a radial tapping switch, ammeter and accumulator, as shown in Fig. 12, the battery being tapped off to the radial switch at 4, 6, 8, 10 and 12 volts. The ammeter was of 0 to 2 amp. range (A).

The output to the tank is fed through the main ammeter A_2, and the variable resistance R, and the system of operating the plant was as follows.

The tapping switch was set to whatever tank-voltage was required, and the generator started up. The main resistance R, and shunt resistance S were then manipulated so that the correct plating current was shown by the ammeter A_2, and a very small charge, say $\frac{1}{4}$ amp., was shown by the ammeter A.

This was the neatest, simplest and most efficient home-made plant that the author has ever seen, and has proved to be absoluetly trouble-free over a period of two years.

There is little else to be said about D.C. mains, except that in all cases where no voltage variation can be obtained from the dynamo, the control resistance, whilst taking the same general pattern as the one described earlier in this chapter, must be made on a larger scale; it being the sole controlling element, it will need to have the number of sections increased, so that a total resistance of $V \times 2$ ohms is obtained, where V is the output voltage of the dynamo. The sections, also, will need to be more robust, and should all (except perhaps the last 4 ohms on the maximum resistance side) be capable of passing about 6 amps. The low resistance coils at the extremity should be capable of passing the greatest current likely to be needed, say 10 amps.

Finally, although stress has been laid on the dynamo and electric motor as a source of plating current from D.C. mains, there is, of course, no objection to any type of engine for driving the dynamo, provided it is capable of

26

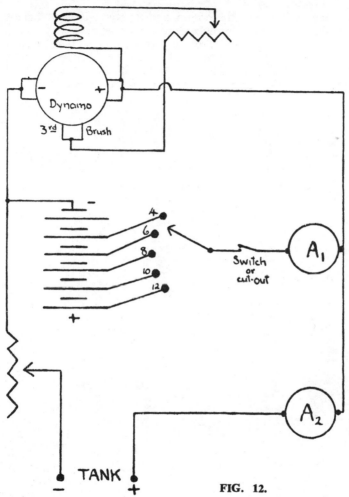

FIG. 12.

Dynamo and Battery Set. Ammeter A_2 should be fitted with variable shunts as in Figs. 10 and 11.

27

EPA–C

doing so for a lengthy period, and is fitted with an efficient governor to regulate the speed.

Those readers who are not fortunate enough to have a supply of current laid on would be well advised to consider whether this, in the long run, would not be more economical than batteries, in spite of the higher initial cost.

3 *THE PLATING TANK*

THE tank or vat in which the actual plating is done is, theoretically, only a means to an end. That is, it takes no actual part in the process except to hold the electrolyte, and allow immersion of the base to be plated and the anodes, and as most solutions are used at an elevated temperature, it must be able to withstand heat.

Unfortunately in practice most of the solutions to be used are of a corrosive nature, and consequently the material of which the tank is composed must be chosen with this fact in mind. A brief consideration of the various types of electrolyte, as the plating solutions are called, will give an indication of the most suitable materials.

Broadly speaking, there are two main classes of electrolyte : the acid class and the alkaline class. They can be tabulated as shown in Table 2, which indicates the materials which the various solutions attack, and which therefore must not be allowed to come in contact with them.

A study of this table will show that there are only three materials suitable for holding all types of electrolyte, i.e., glass, glazed earthenware, and stoved enamel. Glass is not very satisfactory, owing to the fact, as mentioned above, that most electrolytes are used at higher temperatures than the room, and heating arrangements for a glass tank would present numerous difficulties, because of its fragile nature. Glazed earthenware, in order to be sufficiently robust,

TABLE 2

ACTION OF ELECTROLYTES ON VARIOUS MATERIALS, AT WORKING TEMPERATURES

(The ideal material for plating tanks for the various electrolytes are marked with an asterisk)

ELECTROLYTES	No Action On	Negligible Action On	Strong Action On
A. Acid Types:			
No. 1 Copper	*Copper *Glass *Glazed earthenware	*Cellulose enamel *Stove enamel Lead	Zinc, Iron Tin, Brass
Chromium	Glass Glazed earthenware	Cellulose enamel *Stove enamel Lead, Aluminium	Copper, Iron Zinc, Tin Brass
Nickel	Copper *Glass *Glazed earthenware *Stove enamel Cellulose, Lead Iron, Brass	Zinc Aluminium	
B. Alkaline Types: Silver Cupro-Nickel No. 2 Copper No. 2 Zinc	Glass Glazed earthenware	Copper Lead *Stove enamel	Zinc Aluminium Brass Cellulose enamel

would need to be very heavy, and here again heating difficulties are met. Both of these types of tank could be heated only by an immersion heater, and the temperature would have to be raised very slowly, in order to avoid the danger of breaking. Nevertheless, in spite of these drawbacks, glass and earthenware have a very strong recommendation in the fact that they are the only materials completely unaffected by all types of electrolyte, and if a low-power immersion heater can be installed and adequate agitation provided, a glass or glazed earthenware tank is highly satisfactory.

A further glance at Table 2 will show that the only type of tank bearing an asterisk in all cases is the stove-enamelled one, and by this is meant an iron tank with a vitreous finish, applied at a red heat. This finish, being in reality a kind of glass, is almost completely unaffected by all electrolytes.

Such tanks can easily be adapted from large domestic breadbins, and here again, these are precisely what the author used to make his tanks. A complete description of how one of these tanks was fitted (they are all exactly alike) will no doubt furnish the beginner with all the information needed in order to make a really first class tank.

When purchasing a breadbin, it should be examined carefully for cracks and chips in the internal glazing, and, if electric heating is to be fitted, should have a flat, not a ribbed bottom. Only one such tank will actually be required for plating, but it should be borne in mind that if the various solutions are to be stored when not in use, then a further two will be required. This is more fully explained in Chapter 5, where the actual solutions are discussed, but it can be mentioned here that if oil drums are to be used for storing the solutions, they are not satisfactory either for the chromium or silver, and it is suggested that these two solutions be kept always in an enamelled breadbin. The remainder of the solutions can be kept in tinned iron-ware, such as cleaned oil drums, but only when cold. The action of the various solutions when cold is sufficiently small to be neglected, but tinned iron is quite unsuitable for plating tanks when the electrolytes are hot.

The preliminary considerations regarding a plating tank, apart from the material of which it is constructed, are concerned with : 1. Size and shape of base article. 2. Position, distance and number of anodes. 3. Fitting of anode- and cathode-bars. 4. Heating arrangements. 5. Agitation. 6. Fume clearance. These items will be dealt with, now, in the above order.

Items 1 and 2 are concerned only with the size and shape of the tank, and it is quite possible to obtain a breadbin of approximately 11 in × 13 in. × 12 in. deep, which is an excellent size to accommodate a base of the maximum dimensions, as laid down in Chapter 1. Such a tank will hold, when filled to within half an inch of the rim, about six gallons, and the quantities of chemicals to be dissolved in this amount of water can easily be calculated from the formulæ given in Chapter 5, as these quantities are all expressed as ounces per gallon (oz./gall.) and require only to be multiplied by six.

If the tank obtained differs in dimensions from the above, its cubic capacity in gallons can be calculated from the formula

$$G = \frac{L \times B \times H}{260}$$

where L, B and H are dimensions of tank in inches, and G is the capacity in gallons.

It might be noted here that for nickel and acid copper electrolytes, which are used cold, an excellent tank can be made from an aquarium tank. Those constructed of steel angle-iron and glazed with non-oily " putty " are best, and can be had in all sizes. They are not very suitable for warm electrolytes, as the alternative heating and cooling affects the putty.

32

ELECTRODE BARS

The fitting of anode and cathode bars can be discussed next. The requirements for small-scale work are quite simple, and consist of two conductors from which the anodes are suspended, and one conductor from which the cathode is suspended. These bars are preferably of copper,

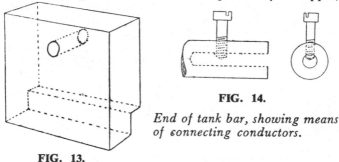

FIG. 14.

End of tank bar, showing means of connecting conductors.

FIG. 13.

Wooden Blocks to support and insulate the tank bars.

but brass is quite an efficient substitute, and is much easier to work. The author's tanks are all fitted with brass bars, and the only attention needed is an occasional cleaning with sand-paper. The bars are of $\frac{3}{8}$ in. round rod, carried in wood holders which rest on the sides of the tank, and may be constructed as in Figs. 13 and 14.

The bars are cut 4 in. longer than the shorter dimension of the tank top, and a $\frac{3}{16}$ in. central hole in each end is drilled to a depth of 1 in. Half an inch from the end, a hole is drilled diametrically and tapped $\frac{1}{8}$ in. for a set-screw to bind the connecting wire in the hole. The wooden blocks, cut to the shape shown, are then drilled a tight pushing fit over the brass rods, and driven on. Their positions are adjusted so that the grooves cut in the bottom of each rest nicely over the rim of the tank, one over each side. Three such bars and supports are made and are fitted with movable contact bosses, as shown in Fig. 15, cut from $\frac{1}{4}$ in. sheet brass. Alternatively, wooden runners

33

may be fitted along front and back of the tank, each being drilled with numerous holes to accommodate as many bars as may be required.

The hole is drilled $\frac{3}{8}$ in. clearance, and the binding screw is $\frac{1}{8}$ in. Whit., the connecting screw $\frac{3}{16}$ in. Whit., both in brass. Three such bosses are made for each bar, making nine in all. This number, of course, could be increased if many articles were to be plated at once.

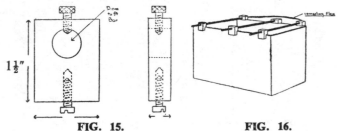

FIG. 15.   **FIG. 16.**

Details of Connecting Bosses.   The tank with bars in place.

Fig. 16 gives an idea of how the tank appears when fitted with its conductors. It will be noted that the two outer, or anode bars, are shown connected together, as this is the usual arrangement for plating all sides of an article at once. The centre bar is the one from which the base is suspended, and the wire connecting the two anode bars is a heavy flex, which, whilst carrying the current quite satisfactorily, is sufficiently flexible to enable the bars to be moved about.

Having got as far as providing the tank with connecting bars, the next thing is to make some arrangement whereby the contents can be heated. Some metals can be deposited from electrolytes in the cold, i.e., at room temperature. Others need varying degrees of temperature, in order to give best results, and it is advisable, therefore, to provide sufficient heating facilities to raise the temperature nearly to boiling point, if required.

There are numerous ways of effecting this, perhaps the

easiest being to stand the plating tank on a gas ring, and raise the contents to the desired temperature, turning the gas down to a low flame, to maintain that temperature. This is by far the best and most convenient means of heating the tank, there being only one danger to watch out for, and that is, not to apply a very fierce flame to the bottom of the tank, as the vitreous enamel tends to crack.

If no gas is available in the workshop, there remains the alternative of utilising electricity, and this can be done in one of two ways. Either by contact-heating from the bottom, or by an immersion heater.

Heating by the latter means has much to recommend it, but there are also several drawbacks. Chief amongst these is the actual space in the tank taken up by the heater element, and in small tanks of this size there really is not the room to spare. If such a heater is fitted, it must be very small, and consequently, the operator must be prepared to wait a long time for the electrolyte to heat up. Any heater of this type must be inserted from the top of the tank, to rest upright in one corner, and should be removed, rinsed and dried, when not actually in use. The outer sheath MUST be either of copper, silica or glass and, if of copper, must be solid-drawn, as soldered or brazed joints are not permissible, solder being attacked by most of the hot electrolytes. The glass heaters sold for tropical fish-tanks are quite suitable for heating small installations, but powers exceeding 150 watts are unusual in this type of heater.

The author's tanks are fitted with external heater elements at the bottom, and the following details of the fitting of one of these should enable it to be carried out by any reader who has the necessary equipment in his work shop.

First of all, a heater element is obtained, of the kind used in a large electric iron, known generally as a " tailor's goose." These elements can be obtained at 1,000 or 1,200 watts rating, and this will heat a six gallon tank to boiling point in a reasonable time.

The vitreous enamel is next removed from the centre

portion of the outside of the tank bottom, and this is best done by lightly grinding with a good, sharp carborundum wheel, of a medium grade, taking care not to apply too much pressure, or the enamel will chip off the inside. When an area sufficient to accommodate the heater element has been cleaned, an iron plate is cut, from a piece of $\frac{1}{4}$ in. sheet, of a size and shape to completely cover the element, and leaving an overlap of $\frac{1}{2}$ in. all round. In the centre of the square end, it is cut away to allow the two brass connecting lugs to be brought out (see Fig. 17), and, of course, enough metal must be removed to ensure that these lugs cannot touch the plate at any point. At each corner of this end of the plate, a hole is drilled and tapped $\frac{3}{16}$ in. and a length of steel or brass thread ·screwed in. This should be flush on one side, and project about 1 in. on the other, and is secured with a lock-nut on the projecting side. Two nuts are now run on to these threads to within $\frac{1}{2}$ in. of the lock-nuts, and over these is fitted a strip of $\frac{1}{8}$ in. × $\frac{1}{2}$ in. iron or brass, suitably drilled at each end. Near the centre of this strip are two $\frac{1}{4}$ in. holes, drilled in such positions that the two brass connecting lugs can be bent up and secured to screws passing through them. This terminal strip is fixed by a pair of nuts at each end, and the excess thread cut off.

The next step is to lay the element on this iron plate, and mark out the position of two holes for fixing bolts. These will lie along the centre line, and will pass through the holes or slots in the mica. Having marked these holes, they are drilled $\frac{1}{4}$ in. clear, and the plate is placed centrally on the bottom of the tank, inside. Again the holes are marked through, and the tank is carefully drilled from the inside, so as not to chip the enamel more than necessary. If the drill is taken through slowly, and kept moistened with a drop of turpentine or paraffin oil, very little chipping will occur, and if drilling is done through to a block of brass, taking the bit right through and into the latter, a nice clean hole will be obtained.

The whole is now ready for assembly, as follows.

Two pieces of $\frac{1}{4}$ in. Whit. steel thread are cut, about $1\frac{1}{2}$ in. long, and a thin brass nut is run on to each. This is located so that, when the screw is pushed through the hole in the tank, about $\frac{1}{4}$ in. projects on the inside (the nut being on the outside). Using a blowlamp, and heating up very slowly, these two bolts, with the nuts in place, are sweated into their respective holes with soft solder, making sure that the nuts, as well as the bolts, are thoroughly fixed. On the inside, two steel washers and nuts complete the fixing of the bolts, and at this stage it is advisable to fill the tank with very hot water and leave to stand for half an hour, to make obsolutely certain that there is no leakage.

If all is in order, the tank is turned upside down, and the element placed over the bolts, the mica being carefully cut away, if required, in order to allow the element to pass completely over the nuts and lie flat on the tank. This is then followed by several sheets of asbestos, to give a thickness of at least $\frac{3}{8}$ in., and finally the iron plate, which is very securely fixed by substantial nuts. The first sheet or two of asbestos will need large holes for the nuts.

All that remains is to fix the brass lugs to the terminal strip, and this is done with the aid of $\frac{1}{8}$ in. brass screws, with plenty of mica washers to make sure that there is ample insulation between the contacts and the terminal strip. When correctly adjusted and screwed up, the terminals are each provided with a further pair of washers and a nut for connection to the mains cable. Lastly, one of the $\frac{1}{4}$ in. fixing bolts is fitted with two washers and a nut, to provide an "earth" connection to the cable, and the other bolt is cut off flush.

Earthing of the tank in this way is, of course, absolutely essential, and under no circumstances should any connection of the heater to the mains be made without making sure that there is plenty of liquid in the tank.

The small area of bare iron or steel provided by the nuts and washers of the fixing bolts on the inside of the tank can quite safely be ignored, but for those readers who are very particular about protecting them from rust, it can be suggested that the first exercise in copper plating could very easily be the plating of these parts. They are scrubbed with a nail brush, a little pumice powder and a good deal

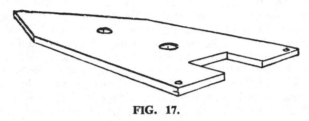

FIG. 17.

Approximate shape of iron clamping-plate for the heating element.

of soap, and the tank is then partly filled with No. 2 copper solution. The tank itself is connected to the cathode bar via the " earthing " terminal underneath, and a small copper anode is immersed in the solution. A very minute current is passed, say, as much as a 12 volt 3 watt side-lamp bulb will pass on the 4 volt tapping, and plating is carried on for ten minutes. The No. 2 copper is then poured out, the tank rinsed with water, and refilled with copper solution No. 1, plating being resumed for a further period of about forty minutes. This will give a robust copper plate, not only to the nuts, etc., but also to any parts of the iron tank left exposed by accidental chipping of the enamel.

A shallow wooden frame should now be made, on which the tank can stand, thus preventing any damage to, or shorting of the connections underneath. A small portion of this will have to be cut away to allow the cable to pass through, and it is essential that this latter be of first quality rubber-covered, to guard against access of any spilled solution to the actual wires.

AGITATION

The final attachment to the tank itself concerns agitation, or stirring of the contents, and perhaps a word in explanation of the need for this will not be out of place.

When a current is passed through an electrolyte containing salts of a metal, the metal itself comes out of the solution, and adheres to the cathode. At the same time, if anodes of the same metal be used, a roughly equivalent amount of metal dissolves from the anodes and passes into solution in the form of salts. If the solution is allowed to remain perfectly at rest, it will be seen that there is a tendency for the electrolyte to become deficient in metal in the region surrounding the cathode, and to become abnormally enriched in the vicinity of the anodes, the nett result being that deposition of metal is retarded, and gassing encouraged. There are also many other important effects on the quality of the plate, which need not be detailed here, but it can quite readily be understood that if the electrolyte is constantly in a state of motion, it will remain at the same uniform strength throughout, and electroplating will proceed at a uniform and satisfactory rate for an indefinite period.

The author has tried various methods of agitation, such as blowing air under slight pressure into the lower part of the electrolyte (the common commercial practice), and by external combined circulating and heating coils, but has found that, for small tanks, none is quite so satisfactory as a propeller, driven by a tiny electric motor.

The motors used are of the miniature 24-volt ex-R.A.F. type, many of which are available at a few shillings each, at the time of writing. These motors, although intended for working on D.C., are almost all provided with laminated fields, and work very well on 16 to 24 volts A.C., direct from the secondary of the transformer. Those installed by the author have, at times, run continuously on 18 volts A.C. for over twenty-four hours, and have not shown signs of heating up. As they are, moreover, totally enclosed, they are reasonably immune to attack by acid or alkaline spray, a little of which is always present near the tank, when in use. In the author's plant, a separate winding

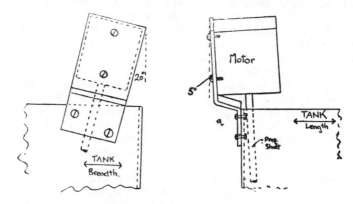

FIG. 18.

Method of setting up the agitator-motor. The propeller should just clear the bottom and the side of the tank.

is provided on the transformer to supply these motors, and the amount of current taken by three working at once is quite negligible.

Fig. 18 shows how such an agitator can be fitted up, and is almost self-explanatory. The propeller-shaft, in the author's case, is a length of stainless steel, which happened to be handy at the time, but it is suggested that a length of copper tubing, of fairly stout gauge, would serve equally well. The driving end of the motor spindle is usually 4 mm., and a piece of tubing which is a very tight fit on this should not be hard to find. This end of the tube is split for about $\frac{1}{4}$ in. and a boss with grub-screw fitted over. The grub-screw should be sunk, to minimise centrifugal force.

At the other end of the shaft, a very small propeller is soldered or screwed on, and this can be made quite simply out of a scrap of 24 or 26 gauge copper sheet. The maximum diameter of the propeller need not be more than about an inch, and the pitch of the blades is found by

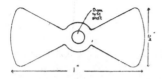

FIG. 19.

The propeller.

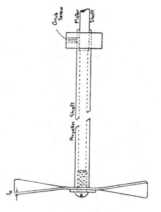

FIG. 20.

Diagram of propeller and shaft. Showing one method of connecting to motor.

trial and error. The object is to arrange the speed of the motor to be between, say, 600 and 1,000 revs./min., and the pitch of the propeller is adjusted so that a reasonable circulation of the electrolyte is obtained. The speed of circulation should be about 3 or 4 seconds per circuit of the tank, as ascertained by allowing a few scraps of paper to drift in it. The speed of the motor is best kept within the above limits, as no lower bearing is provided for the propeller shaft, and high speeds would " pick up " any slight wobble in the shaft, and throw it completely out of line.

Figs. 19 and 20 show shape of propeller.

To those readers who have some experience of electroplating, the foregoing descriptions of the various items of equipment, and the methods of fitting up, will doubtless

41

be of use as suggestions, which are quite capable of modification. On the other hand, the beginner is advised to consider them as definite instructions, which, if faithfully carried out, will produce, if not an ideal plating plant, at least one which will do the job in a very satisfactory manner.

To all readers, however, whether experienced or not, the author would emphasise most strongly the danger arising from fumes.

In the case of chromium plating, very unpleasant fumes are given off by the electrolyte, during passage of current, and, whilst these fumes are dangerous, in that they can cause permanent damage to the skin and mucous membranes of the mouth, throat and lungs, they cannot be said to be lethal. Their presence, moreover, is distinctly noticeable long before they reach a harmful concentration.

In the various electrolytes containing cyanide, however, we meet with a vastly different state of affairs. These electrolytes give off at all times, whether in actual use or not, gases which are very insidious, and infinitely more dangerous than those proceeding from the chromium tank. These gases consist mainly of hydrogen cyanide, which is recognised by its characteristic odour. It is not easy to describe this odour, but it is exactly the same as is noticed when taking the stopper out of a bottle of sodium cyanide. This gas possesses the quality of partially paralysing the sense of smell, and it is quite possible to remain in a room where the concentration of this gas is rising, and yet be under the impression that it is becoming less.

Under actual working conditions, due to the fact that cyanide baths are all worked warm, the evolution of gas is accelerated, and in a closed room can quickly reach dangerous proportions. Hydrogen cyanide is one of the most powerful poisons known, and in lethal concentration it is so rapidly fatal that there is no time for countermeasures to be taken.

All this might appear, perhaps, to be rather dramatic, but the author wishes to emphasise, as strongly as possible,

42

the need for exceptionally free ventilation. If the workshop is in an out-building, all windows and doors must be wide open during the whole of the time that the solutions are exposed, and the operator should not remain in the room longer than is necessary.

If the workshop happens to be in the cellars of a house, as in the author's case, it must be remembered that hydrogen cyanide is a light gas, which rises in air, and if allowed free passage, will concentrate in the bedrooms, etc., at the top of the house.

In such cases, where the plating must necessarily be carried out in a confined space, it is essential to provide for the positive withdrawal of all poisonous fumes. The author solved this problem by fitting a tinplate cowl over the tank, and leading the fumes out through a hole made in a chimney breast, by means of a tinplate shaft, soldered on to the cowl. The cowl and shaft are fixtures, and are arranged in such a position that the tanks can be slid under when about to be used. Owing to the lightness of the gas, and the heat rising from the warm tank, the fumes pass up through the shaft without any actual suction by fan being necessary.

For those who prefer simplicity, and do not mind having to depend on the weather, there is always an alternative in carrying on the whole process out of doors, and the only point to note, in this case, is that the cable carrying the current from the panel to the tank should be as short and as heavy as is reasonably possible, to avoid any considerable voltage drop.

4 *PREPARATION OF THE BASE*

THE importance of this subject has already been stressed in Chapter 1, and a brief explanation as to why the preparation is so important will serve to give further emphasis to this matter.

All the commonly used metals, such as iron, steel, copper, brass, zinc, lead and alloys, and tin are subject to oxidation. This means that, when any of these metals or alloys are rendered perfectly clean, and then left exposed to the atmosphere, a film of metallic oxide develops on the surface. The rate at which this film is formed depends on many factors, chief among which are the nature of the metal, the temperature, and the amount of moisture in the air.

For a variety of reasons, any electro-plate which is deposited on a metallic base must be in actual physical contact with the base, as, if a film of oxide amounting, in some cases, to as little as 0.00001 in. is allowed to come between, the electro-plate will not adhere to the base.

Similarly, if the surface of a piece of metal is rendered perfectly clean, and then touched by hand, a very minute trace of grease is left upon it, and, while this will not usually prevent the deposition of plate, it will invariably cause the plate to " lift " at the point in question.

It will thus be apparent that there are two points to be borne in mind in the preparation of a base for plating. Firstly, the metal must be stripped of all oxide, and secondly, having been rendered perfectly clean, it must not be touched by hand until the plating is completed.

44

It is not the purpose of this book to enter into a discussion of the machining of castings, etc., it being assumed that whatever articles the reader might wish to plate will be in a more or less finished condition. Any castings which have not been machined over the entire surface should be made reasonably smooth, either by filing or grinding, before attempting to prepare for plating.

The actual preparation is carried out in some, or all of the following stages :

1. Smoothing, i.e., grinding, smooth-filing, sand-papering.

2. Partial polishing, i.e., buffing with very fine emery and grease, or with coarse polishing compo.

3. Chemical cleaning, by immersion in either acid or alkaline baths.

4. De-greasing.

5. Electrolytic cleaning, in which the base is immersed in suitable solutions, and a current passed.

6. Etching, to provide a very fine matt surface.

7. In the case of chromium plating only, a very high polish of the under-coating, prior to final plating.

8. For silver-plate on copper or brass, the quick-silvering process.

9. Iron dipping.

As different bases require differing treatments, it is proposed, for the sake of clearness, to describe the various processes outlined above, in detail, and then to tabulate the various bases (in Table 3), to show which processes they require, and in what order they are to be carried out.

Process 1 (Smoothing)

This is purely a matter of removing all indentations and excrescences from the base, and reducing it to a reasonably smooth surface. The file, grindstone, and emery or sand-paper all have their uses, and, with a little patience, quite a good surface can be obtained. At this stage, the base

45

should present a clean appearance with none but the very lightest scratches visible.

Process 2 (Partial polishing)

A light polishing with grade " o " emery, bonded with grease, and used on a calico wheel. The author finds this useful for cast iron and hard steels, where sandpapering is too tedious. It is not required for softer bases.

Process 3a (Alkaline dip)

The base is immersed in the following solution, at room temperature, for five minutes :

Pure Caustic Soda	...	6 ozs.
Sodium Carbonate	...	6 ,,
Sodium Cyanide, 130%	...	4 ,,
Water to ...	... ...	1 gallon

Process 3b (Acid dip)

The base is immersed in the following solution, at room temperature, for three minutes :

Sulphuric Acid, pure	...	50 ozs. (fluid)
Hydrochloric Acid, pure	...	2 ,, ,,
Nitric Acid, pure	...	3 ,, ,,
Water ...	... ...	60 ,, ,,

Warning

In preparing this solution, the following routine must be observed, without deviation :

Mix the water, nitric acid and hydrochloric acid. Add one quarter of the sulphuric acid, and cool thoroughly. When cold, add a further quarter of the sulphuric acid, and again cool. When quite cold, add the remainder of the acid, and cool. The solution should be kept in glass containers with a glass stopper.

Process 4 (Degreasing)

The base is carefully swabbed over with a small piece of clean rag, saturated with carbon tetrachloride, changing the rag at least three times. This is followed by scrubbing with hot water and household detergent, using a clean nail-brush.

Process 5 (Electrolytic cleaning)

The base is connected to the cathode bar and immersed in the following solution, at 70° Centigrade, and a current of 2 amps. per square decimetre is passed for five minutes, using a steel plate for anode :

Sodium Cyanide	4	ozs.
Sodium Carbonate ...	6	,,
•Sodium Tribasic Phosphate	3	,,
Caustic Soda	1	oz.
Water to	1	gallon

Process 6 (Etching)

6a For non-ferrous metals

The base is made ANODE, and immersed in the following solution, at 70° Centigrade, and a current of 2 amps. per square decimeter is passed for three minutes, using steel or copper cathodes :

Citric Acid	8	ozs.
Water to make	½	gallon

Dissolve, and add strong ammonia solution, until the liquid just smells of ammonia, constant stirring being necessary whilst neutralising.

Then add :

Citric Acid	3	ozs.
Water to make	1	gallon

6b For cast and wrought iron and steel

The process is identical with the above, except that the following solution is substituted :

Sulphuric Acid, pure ...	5	ozs.
Hydrochloric Acid, pure...	5	,,
Water to make	1	gallon

The cathodes should be copper, and current should be passed for five minutes.

Process 7

The high polish required prior to chromium-plating cannot be produced on cast or wrought iron, and these metals are therefore plated with nickel before chroming. The degree of polish obtained is reproduced exactly by the final chromium, and therefore must be as perfect as possible. The author has found that the red polishing " soap " as it is commonly called is the best medium for producing a mirror surface, either on copper, brass or nickel, and it is a good plan, after polishing thus, to give a very light rub over with a little precipitated chalk and water, on a piece of cotton wool.

Process 8   (Quick-silvering)

This process is used solely to prepare copper, brass and nickel bases for silver plating, and its function is to prevent the silver from lifting. For some unknown reason, silver plated direct on to these metals does not adhere, and a thin film of mercury interposed ensures perfect adhesion.

The process consists in swabbing the base with small pieces of very clean rag or cotton wool, liberally soaked in the following solution, rinsing very rapidly in clean water, and then immediately proceeding with the silver plating. The delay should not be more than ten minutes, and during this time the prepared base must be kept under clean water.

Quick-silvering Solution

Perchloride of Mercury ...	$\frac{1}{2}$ oz.
Sodium Cyanide, 130% ...	2 ozs.
Water to make	$\frac{1}{4}$ gallon

This solution is used cold, and a good pressure is applied with the rag, to distribute the mercury well over the surface. Half the above quantity will cover many square feet, and the solution will keep indefinitely in glass, but must not be allowed in contact with metal during storage.

TABLE 3

PROCESS

BASE	1	2	3a	3b	4	5	6a	6b	7	8	9	Rinse in Water
Cast iron for all plates	A	B			C	E			See Chapter on Chromium Plating			D, F
Wro't iron and steel for all plates	A	B	G		C			E				D, F, H
Copper and brass for silver plate	A		B							D		C
Copper for nickel plate	A			D								C, E
Brass for copper plate	A			B			B					C
Lead for copper and silver ...	A		B									C
Lead alloys for copper, silver, brass and zinc	A					B						C
Tinplate and pewter	A					B						C
Zinc for all plates	A					B	D					C, E; C, E
Aluminium for all plates ...	A	B	C								E	D, F

49

Process 9 (Iron dip)

This is used only on aluminium bases, and serves to give a coating of iron to the base, which can then be coppered in the No. 2 copper bath. There must be no delay between carrying out this iron dip, and the initial coppering.

The process consists of boiling the base in the following solution for five minutes, and should be carried out in an enamel pan.

Solution of Ferric Chloride, B.P.	...	...	...	8 ozs.
Hydrochloric Acid		...		2 ,,
Water to make ...		...		1 gallon

PROCEDURES WITH VARIOUS BASES

Table 3 gives the correct sequence of operations for preparing various bases for various plates, and to read it, take the appropriate base in the left-hand column, and read off the sequence of operations alphabetically. The processes are referred to in the table by the numbers which they have been given in the foregoing paragraphs.

BLOCKING-OUT

The final process of preparation of the base to be described is a means of preventing deposition of plate on certain parts of the base, where this is desirable. Blocking-out is carried out for several reasons, chief of which are firstly, for decorative purposes, where contrasting plates are required, and secondly, for reasons of economy when depositing expensive metals such as gold and silver. For example, the reverse side of a reflector need not be silvered, and blocking-out will prevent wasteful plating of this metal, which is quite expensive.

The process is quite simple, and is carried out immediately after completion of other preparatory processes, so that the work may be introduced to the plating tank without further handling. Various varnishes are in use as blocking agents, but cellulose lacquer appears to be the most satisfactory. The clear kind is perhaps best, and one should choose a "brushing" quality rather than one intended for spray application, as the latter has very poor adhesion to bright metal surfaces. A single coat is sufficient, using a soft brush, and of course taking extreme care that none is allowed to come in contact with parts requiring plating. After the plating is complete, the lacquer may be removed with cellulose thinners on a rag..

At least an hour's drying-time should be allowed after lacquering before plating is commenced.

CONNECTING THE BASE TO CATHODE BAR

Little need be written about this subject, as it is such a simple matter, but it is surprising to find it completely neglected in so many text-books.

There are four commonly-used methods, and the choice depends on the base to be plated.

1. Soft-soldering.
2. Nut and bolt.
3. Spring clip.
4. Saddle wires.

Of these, soft-soldered connections, where possible, are by far the most satisfactory, and should be adopted in all cases where a little space can be found which need not be plated, or which will not be seen in the finished object. Examples of this are :

Inside of teapot; rear side of reflector (close to rim); inside cup of candlestick; under extreme tip of leg of small article, such as sugar-bowl.

The nut and bolt is a very good means of making connection, if it can conveniently be fitted. Its use demands the existence of suitable holes in the base, and examples of its application are the centre holes in flywheels, locomotive wheels, etc., existing construction holes in parts of

51

models, which are often plated in portions, before assembly, and, of course, the screw can be fitted into any existing tapped holes, the nut then being unnecessary.

When neither of the above methods can be adopted, there remains either the " crocodile " clip, of the type used for connection of charging leads to car batteries, or, finally, the wire sling. The clip needs nothing by way of explanation, but it is essential when using it to clip it on to a portion of the work not requiring a plate, or if this is not feasible, to move its position several times during the plating operation. Otherwise, little or no plate will be deposited on that part of the base which is " shaded " by the clip.

In cases where the whole of the base is required to receive a uniform plate, the wire sling is the most satisfactory connector. This varies in shape and size according to the particular base to be held, and a few examples are given in Fig. 21.

The sling wire is 16 to 18 s.w.g. copper, bared and thoroughly cleaned along its whole length, by scraping or sandpapering. A most important point when using it, is to ensure that the actual points of contact are moved periodically, as mentioned in regard to the clip, above. The position must be shifted at least four or five times during plating, and this can usually be accomplished without taking the job out of the tank, simply by pushing the wire a little out of place with the aid of a couple of clean wooden skewers. Current need not be switched off, and very little movement is necessary, a $\frac{1}{4}$ in. being ample as a rule.

A point well worth mention at this stage, concerns the actual thickness of metal deposited on the various surfaces of the base, and, without entering into any technical details regarding " throwing power " of various solution, it can be said, quite briefly, that more metal is deposited on the parts of the base nearest to the anodes than on those parts more remote. All this boils down to the fact that it is a great practical advantage, wherever possible, to give each

52

side or face of the base an equal chance, by turning round occasionally during deposition. For example, a cube, as shown in Fig. 22, which required thirty minutes' plating, would best be treated by allowing ten minutes in each of the three positions shown. The principle applies to all work, and to all electrolytes, but most particularly to the acid types, i.e., No. 1 copper, No. 1 nickel, and chromium. Whatever the actual shape of base, the operator should use his initiative in ensuring that all surfaces each have their turn to be exposed to the anodes.

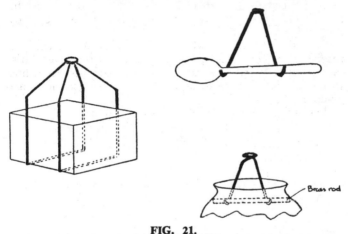

FIG. 21.

Wire Slings, various types.

The author has found the plating of small spoons to be one of the most awkward problems in respect of connection to the cathode bar.

The ideal position for a spoon is on its edge, that is, suspended horizontally, with the blade in a vertical plane, as shown in Fig. 21, with the front (concave) side twice as far from its facing anode as the back (convex) side. The reason for this is the need to have a greater weight of metal deposited on the back, as this side is subject to much more wear than the front. The reader will find it a good exercise in manipulation to arrange a wire sling, of the pattern sketched, which, whilst holding the spoon in the correct position, will yet allow sufficient play for the spoon to be moved about, thus producing a uniform plate. As a matter of fact, when the beginner has advanced so far as to be able to produce a fine, even plate of silver on half a dozen teaspoons, he can rest assured that he has progressed well beyond the " novice " stage.

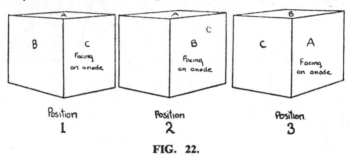

FIG. 22.

Position of base, showing the shifting around necessary to produce an even plate.

One final point, before proceeding to a consideration of the various electrolytes. This concerns the high price of silver, and the fact that a considerable quantity of this metal is deposited on the sling-wires and connectors generally.

Rather than allow this silver to be wasted, the following process of recovery is carried out. By recovery, is meant, of course, the putting back of this silver into the solution, where it is available for depositing again. The various wires are bunched together, and suspended from the anode bar. One of the usual anodes is taken from the normal

54

position and connected to the cathode bar, the other anode being removed from the tank altogether. A small current, equivalent to 1/30th amp. per inch length of the wires, is passed, until all the silver is dissolved off them. A little copper will also pass into solution, but this does not matter, as it will not be deposited under working conditions of the bath.

During this process, an amount of silver almost equal to that dissolved will be deposited on the anode (now cathode), but this will again pass into solution when the anode resumes its normal role. The whole of the silver from the sling-wires is thus recovered.

5   *ELECTROLYTES*

Warning. Many of the following solutions contain sodium cyanide. This is a most dangerous poison, and must at all times be handled with extreme care. Under no circumstances should acid be allowed. in contact with it, or with solutions containing it, and it must be stored out of reach of children and any persons not familiar with its dangerous nature. All solutions of cyanide give off, when exposed to the air, prussic acid gas. On no account should this be inhaled as it is highly poisonous, even in very small quantities. This evolution of gas takes place much more rapidly when the solutions are warm, and when they are being used electrolytically to deposit metals. Whenever electro-plating is being carried on with cyanide-containing solutions, either the fumes must be carried away by means of a suitable cowl and duct, or else the whole operation must be conducted either out in the open or in a well-ventilated shed. In the latter case, the operator must not remain in the shed for longer than is absolutely necessary to make required connections and adjustments to the apparatus. The reader is referred to previous notes on this subject, in the latter part of Chapter 3.

We arrive, now, at the preparation of the various electrolytes, which are the solutions of metallic salts from which the required plates are deposited by the passage of current.

There are one or two points to be strongly emphasised, chief amongst them being the fact that it is quite useless to prepare any of the following solutions from impure materials. All chemicals used must be of the highest grade, and exactly as specified. For example, if the formula given stipulates " Sodium cyanide, 130%," this salt only must be used; it would be quite futile to employ potassium cyanide, which might vary in cyanide content between 20% and 98%.

Similarly, although the exact composition of some of the electrolytes is not critical, in others it is extremely so, and the reader who has little or no experience should therefore aim at absolute accuracy in all of them. Pure chemicals are available through chemical trade channels, but these are not readily open to the average person, and the author has always found much satisfaction in the high standard of chemicals obtained through the retail pharmaceutical trade. Few pharmacists have a stock of all the chemicals required by the amateur electro-plater, but most of them will willingly obtain what is required at a few days' notice.

Cyanides, and the salts of mercury, are controlled by the Pharmacy and Poisons Acts, and, in addition to the purchaser having to sign the poison book, he must either be known to the vendor, or else be introduced by a mutual acquaintance. Such is the law, and it applies in so far as the subject of this book is concerned, only to cyanides, and mercuric chloride. It should not be assumed, however, that these are the only poisonous substances used, and it is a good plan to regard all chemicals as poisonous, and treat them with due care. Any solutions which are spilled should be immediately mopped up, and under no circumstances should any of the various cyanide solutions be allowed in contact with any liquid containing acid, as this gives rise to copious evolution of the intensely poisonous prussic acid gas.

To proceed, then, to the actual formulæ for electrolytes, it will be noticed that these are given titles, such as No. 1 copper, and so on. These titles are referred to in the tables of procedure later in this chapter. Where no individual instructions for making up are given, the various substances are dissolved in three-quarters of the amount of water, in the order given, and the final solution adjusted to correct volume with water. Unless otherwise stated, cold water must be used.

After each formula are notes regarding the storage, when not in use, the correct plating temperature, current density, and any individual peculiarities. The current densities are expressed as so many amps. per square decimetre. The number of square decimetres is obtained by dividing the surface area in square inches by 15.7 (for all practical purpose, 16).

NO. 1 COPPER

	oz./gal.
Copper Sulphate, crystals	10
Sulphuric Acid	1 (fluid)
Sodium Sulphocarbolate	$\frac{1}{8}$

Method: Carefully pour the acid into three times its volume of COLD water, stir well. This generates much heat, and whilst still hot, add the sodium sulphocarbolate. Boil gently in an enamelled iron pan for three minutes, pour into three-quarters of the total amount of water, dissolve the copper sulphate, and make up to volume with water.

Plating Temperature: Cold (12 to 20° Centigrade).

Plating Current: 1 to $1\frac{1}{2}$ amps. per square decimetre.

Anodes: Sheet copper, twice area of base.

Time: To deposit 0.0001 in. plate at $1\frac{1}{2}$ amps., eight minutes.
Solution to be stored in glass or glazed containers.

Correct Appearance of Cathode during Plating: Uniform pink, semi-lustrous.

Signs of Excessive Current: Brownish, rough patches, particularly on projections and edges.

NO. 2 COPPER

	oz./gal.
Copper Sulphate, crystals	2½
Sodium Cyanide, 130%	4
Sodium Bisulphite ...	3
Sodium Carbonate ...	2
Caustic Soda	¼

WARNING. During the preparation of all solutions containing copper sulphate and cyanide, poisonous Cyanogen gas is evolved. This is a very dangerous gas, and all such preparations MUST be done outdoors or in an exceptionally well-ventilated place. On no account should these fumes be inhaled.

Method :· Dissolve the copper sulphate in half the water, and the sodium cyanide in a quarter. With constant stirring, pour the copper solution into the cyanide solution, in successive small amounts. The resultant liquid should be brownish, with no trace of green. If any greenish tint is visible, add a little more cyanide, dissolved in water.

Dissolve the caustic soda, sodium carbonate, and bisulphate, in that order, and make up the volume with water.

Plating Temperature : 55°C. 40° if blocked out with wax.

Current Density ; ¼ amp. per square decimetre.

Anodes : Sheet copper, as for copper No. 1.

Time : To deposit 0.0001 in., forty-five minutes, approx.

Correct Appearance of Cathode during Plating : Clean pink, a little more red than with acid copper, somewhat lustrous at first, and becoming satiny, then matt.

Faults during Plating :

A. Rough brown powdery patches on cathode, with excessive gassing, indicates too heavy current.

B. Black or brown patches on cathode, with little gassing, indicates insufficient agitation.

59

c. Excessive green powder on anodes, or suspended in the electrolyte indicates insufficient cyanide in the solution.

Solution may be stored in any suitable container, except galvanised or other zinc ware.

NO. 1 NICKEL

	oz./gal.
Nickel Sulphate, crystals	12
Epsom Salts	4
Boric Acid	1½
Ammonium Chloride ...	1
Glucose	1

Method: Boil one-eighth of water with the boric acid, in enamelled pan, till dissolved; pour into three-quarters remainder of water, at about 80° Centigrade. Dissolve nickel, epsom and ammonium salts, and finally the glucose. Allow to become quite cold, and decant, or strain through fine linen.

The glucose is an interesting ingredient, in that it plays no part in the chemical changes which occur during plating, but for some obscure reason, it considerably brightens the deposit, with consequent saving in the final polishing. It is known as an "addition agent."

Plating Temperature: Cold (15 to 20° Centigrade).

Current Density: ½ amp. per square decimetre.

Anodes: Nickel bars, three-quarter area of cathodes.

Time: To deposit 0.0001 in., thirty minutes.

Solution may be stored in any convenient container, which must be quite clean.

Correct Appearance of Cathode during Plating: Silvery, tending to become duller after thirty minutes.

Faults during Plating:

A. Dark streaks, usually vertical, indicate insufficient agitation.

B. Rough, grey patches mean excess current.

c. Failure to deposit indicates excess acid. Correction of this is difficult, but try adding successive small quantities of pure borax, say $\frac{1}{4}$ oz. at a time, dissolved in 5 ozs. water. Stir well, and try a test piece of plating after each addition.

Amongst the numerous electrolytes used in plating, this is the only one which has absolutely no cleaning properties, and, therefore, the need for scrupulous cleanliness of the cathode, prior to immersion, is imperative.

NO. 2 NICKEL

(*Cupro-Nickel, or German Silver*)

	oz./gal.
Nickel Sulphate Crystal	4
Copper Sulphate Crystal	12
Sodium Cyanide, 130%	18
Sodium Carbonate ...	6

WARNING. During the preparation of all solutions containing copper sulphate and cyanide, poisonous Cyanogen gas is evolved. This is a very dangerous gas, and all such preparations MUST be done outdoors or in an exceptionally well-ventilated place. On no account should these fumes be inhaled.

Method : Dissolve the nickel and copper salts in half the water. Dissolve the cyanide in one-quarter of the water, and pour the former into the latter stirring constantly. Dissolve the sodium carbonate in the solution, and make up to volume with water.

Plating Temperature : Variable between 40 and 90° Centigrade (see below).

Current Density : $\frac{1}{2}$ amp. per square decimetre.

Anodes : Three copper, one nickel.

Time : To deposit 0.0001 in., twenty-five minutes.

Solution may be stored in any container, except zinc or galvanised ware.

61

Correct Appearance of Cathode during Plating: With rising temperature, from grey, through yellowish grey, to brownish grey.

The final plate can be varied, in this process, according to the following table:

Temperature	Approx. % Nickel	Copper	Appearance (Polished)
40°C.	50	50	White, silvery
50°C.	45	50	„ „
60°C.	41	59	Yellowish „
70°C.	36	64	„
80°C.	32	68	White
90°C.	27	73	White, high lustre

The higher percentage nickel alloys are very much like pure nickel except that they are a little easier to polish.

The medium percentages give a yellowish plate, somewhat similar to the so-called " white gold."

The higher-copper alloys are white, extremely ductile, and take an exceptionally good polish. They are very good indeed for reflecting surfaces, and do not tarnish so easily as pure nickel, any slight tarnish which is formed being much more easily removed than from nickel, with the aid of ordinary household metal polish.

NO. 1 ZINC

	oz./gal.
Zinc Sulphate	6
Sodium Cyanide, 130%	9
Sodium Carbonate ...	2
Caustic Soda	1
Sodium Sulphite ...	3

Prepare exactly as for No. 2 copper, and finally, add:
Potash Alum ...   ...   2 oz./gal.

Plating Temperature: Cool (20 to 25° Centigrade).

Current Density: 1 amp. per square decimetre.

Anodes: Sheet zinc.

Time: To deposit 0.0001 in., fifteen minutes.

Correct Appearance of Cathode during Plating: Bluish, bright deposit, somewhat milky after thirty minutes.

Faults during Plating:

A. Grey or whitish matt deposit indicates excessive current.

B. Heavy coating of white powder on anodes, indicates lack of cyanide in solution.

Solution may be stored in any convenient container,

NO. 2 ZINC
(*White or Hard Brass*)

	oz./gal.
Zinc Sulphate	4
Copper Sulphate	1
Sodium Cyanide, 130%	7
Caustic Soda	$1\frac{1}{2}$
Sodium Sulphite ...	2

WARNING. During the preparation of all solutions containing copper sulphate and cyanide, poisonous Cyanogen gas is evolved. This is a very dangerous gas, and all such preparations MUST be done outdoors or in an exceptionally well-ventilated place. On no account should these fumes be inhaled.

Prepare exactly as No. 2 copper dissolving zinc and copper sulphates together.

Plating Temperature: 20 to 60° Centigrade (see below).

Current Density: $\frac{3}{4}$ amp. per square decimetre.

Anodes: Zinc 2, copper 1.

Time: To deposit 0.0001 in., eighteen minutes.

Correct Appearance of Cathode during Plating: From bright bluish silver to satiny yellow, according to temperature (see below).

Faults during Plating:

A. Dull, heavy matt deposit, or black streaks, indicates excessive current, and/or insufficient agitation.

63

B. Powdery deposits on anodes indicate lack of cyanide in solution.

Temperature	% Zinc	% Copper	Final Plate (Polished)
20°C.	80	20	Blue-white
30°C.	70	30	White
40°C.	60	40	Pale Yellow
50°C.	48	52	Brass
60°C.	35	65	Bronze

The plate laid down at 20° Centigrade looks almost identical with chrome, but is, of course, softer, and tarnishes more easily. All the above plates will take a high polish very easily, and are excellent for protecting iron and steel, in which cases they may be plated direct.

SILVER

(Pure Silver Plate)

	oz./gal.
Silver Nitrate	4
Sodium Cyanide, 130%	6
Sodium Carbonate ...	3
Caustic Soda	1

Method: Dissolve the silver salt in one-quarter of the water, and the cyanide in one-half. Pour the silver solution into the cyanide solution with constant stirring. Dissolve the sodium carbonate and the caustic soda, and make up to volume with water.

Plating Temperature: 30 to 35° Centigrade.

Current Density: $\frac{1}{3}$ amp. per square decimetre.

Anodes: Silver sheet or foil, same area as cathodes, or steel.

Time: To deposit 0.0001 in., fifteen minutes.

Correct Appearance of Cathode during Plating: Pure white or slightly yellowish, flat surface, resembling white velvet.

Faults during Plating :

A. Black or brown patches on cathode indicate excess current and/or insufficient agitation.

B. Failure to deposit indicates exhaustion of silver from electrolyte (this only occurs when using steel anodes).

It should be explained at this stage, that the use of steel anodes causes the silver content of the solution to diminish during plating, as opposed to the use of silver anodes, in which case the silver content of the electrolyte remains constant. The only reason that steel electrodes are suggested is that silver sheet or foil is not so readily obtainable as silver nitrate. If steel anodes are used, it will be necessary to replace the silver in the solution, from time to time, and this is done by dissolving 3 ozs. of silver nitrate in about 10 ozs. water, and adding to the solution with constant stirring. This should be followed by the addition of $1\frac{1}{2}$ ozs. sodium cyanide, dissolved in a small amount of water. These additions to the solution can be made about five times, after which the solution, on final exhaustion of silver, should be discarded and a fresh solution made.

Solution should be stored in non-metallic container, or stove-enamelled bin, to avoid loss of silver by deposition on metal surface.

NO. 2 SILVER

Former editions of this book have given details of a combined silver and copper electrolyte under the above title, which will produce an effect variable between silver and a rich golden colour.

It has been deleted from the present edition because it has been found that, for amateur use, the conditions of plating, together with solution composition, have proved to be too critical. The author has carried out some excellent work with this bath, and little difficulty was experienced. However, it appears from reports made from time to time by would-be users of this electrolyte that it is somewhat erratic. The reasons are probably complex, but are most likely concerned with purity of ingredients and water supply, together with the fact that electrolytically pure copper must be used for anodes. For these reasons, therefore, it has been decided to delete it, but if any reader wishes to experiment, he should consult either of the two previous editions.

Chromium

This is dealt with as a separate subject in a later chapter.

Gold

It may happen that the reader is interested in the gilding or re-gilding of small articles such as jewellery, watch-cases, etc., and in that case, the following description of a simple method of carrying out such plating may be of value.

For small articles, there are two commonly used methods of gold-plating. Firstly, there is the normal method where the base is connected to the cathode bar in a small tank, and plating carried out in the usual way, using an electrolyte containing gold salts in solution, and either a gold or carbon electrode at the anode bar. The second method consists in making a " cell ", which is a type of battery in which the base to be plated constitutes one of the poles of the cell, and generation of current is automatic, and independent of outside supply. This latter method is probably the simpler, but as an electrode of fine gold (24 carat) is required, it will probably be outside the scope of the average amateur. For this reason, the present remarks will be confined to the first method.

Preparation of the base

Such articles as are normally required to be gold plated are usually made of brass or bronze, or occasionally, silver. These bases require no preparation other than an exceptionally good polishing, followed by a thorough de-greasing as described in Chapter 4. They are then slung on a suitable

66

wire and introduced to the plating tank. In the event of steel or iron articles being involved, it is strongly recommended that they be first plated with either copper or nickel, as gold has very poor adhesive properties when applied direct on ferrous bases. In all cases, the most important thing is to obtain a highly polished surface completely free from grease, and, as stressed previously in this book, bases prepared for the tank should *never* be fingered.

The Tank

As only small articles are envisaged, the most suitable tank is an enamelled iron pan, and one holding a pint will most likely be quite big enough. Under no circumstances should a pan of bare metal be used, as gold will deposit on practically all metals *without the passage of current*, and the solution would thus be depleted of its gold content. The reason for not making use of this automatic deposition of gold in the actual plating is that it is not sufficiently adhesive when so applied, and, secondly, such a plate is limited to an extreme thinness, further deposit of gold being prevented when once the article is covered by a very thin layer.

The Electrolyte

The following is the formula for one pint of solution.

Dissolve 30 grains (apothecaries weight) of Gold Chloride in half a pint of lukewarm water. Dissolve one drachm (apoth.) of sodium cynanide (130%) in one ounce of water. Mix the two solutions. Dissolve half a drachm of sodium cyanide and half a drachm of pure caustic soda in one ounce of water and add to the gold solution. Make up to volume with cold water. Store in glass container, in the dark.

Anode

Either stainless steel, say equal to base area, or carbon. The latter is recommended, and may take the form of a rod of graphite removed from the middle of a small torch cell. This should be soaked in several changes of water for a few days to remove the sal-ammoniac and zinc salts with which it will be saturated. The brass cap should not be allowed in contact with the gold solution.

Plating temperature

About 50 to 60 deg. Centigrade gives good results, but the colour of the gold plate depends to some extent on the temperature. Lower temperatures give a redder colour, and higher ones a yellowish tinge.

Current Density

This is in the region of $\frac{1}{4}$ amp. per sq. dm. but as the base will be considerably smaller than one sq. dm. the correct current is best found by trial. It is suggested that somewhere near the correct current will be passed if the transformer tapping is set at 4 volts, and the whole of the resistance is in circuit.

Method

Having prepared the base, and slung it from a thin copper wire, this is connected to the negative supply terminal. The anode is connected to the positive, and the solution placed in the pan and heated to the correct temperature. Next, the anode is lowered into the bath, followed by the base, which is kept moving by hand for about one minute. This will usually give a thick enough plate, and thickness can be estimated by rubbing with a rag soaked in liquid metal polish. If the gold plate resists this treatment for twenty minutes, one might consider that it is thick enough to withstand a fair amount of ordinary wear and tear. Those readers who are equipped with a delicate balance could, of course, obtain an accurate figure by weighing before and after plating. On small articles, however, the weight of gold deposited will be found to be so small as to be almost impossible to weigh.

When using an insoluble anode, as above, the gold is gradually removed from solution, and eventually no deposit will be obtained. When this occurs, the old solution should be discarded, and a fresh one prepared.

The following tables show the sequence of operations for depositing the various plates on various bases, and it is assumed that the preparation of each base has been carried out, as detailed in Table 3, Chapter 4.

TABLE 4

PLATING ON CAST IRON

Final Plate Required	Copper	Zinc	Nickel	Pure Silver	Cupro-Nickel Hard Brass
Process 1 ...	No. 2 Copper 10 minutes	No. 1 Zinc, 40 minutes	No. 2 Copper, 40 minutes	No. 2 Copper, 10 minutes	No. 2 Copper, 10 minutes
Process 2 ...	Rinse	Rinse, dry and polish	Rinse, dry and polish	Rinse	Final plate (without rinsing), 1 hour
Process 3 ...	No. 1 Copper, to required thickness		No. 2 Copper, 10 minutes	No. 1 Copper, 30 minutes	Rinse, dry and polish
Process 4 ...	Rinse, dry and polish		Rinse in 5% Sulph. Acid	Rinse, dry and polish	
Process 5 ...			Rinse in water	Quick-silver	
Process 6 ...			No. 1 Nickel, 1 hour	Silver, 40 minutes	
Process 7 ...			Rinse, dry and polish	Rinse, dry and polish	

TABLE 5

PLATING ON WROUGHT IRON AND STEEL AND ON ALUMINIUM, FOLLOWING THE IRON DIP (Process 9, Chapter 4)

Final Plate Required	Copper	Zinc	Nickel	Pure Silver	Cupro-Nickel Hard Brass
Process 1 ...	As for Cast Iron	As for Cast Iron	No. 1 Nickel, to required thickness	No. 1 Nickel, 10 minutes	Final plate direct, to required thickness
Process 2 ...			Rinse, dry and polish	Rinse and quicksilver	Rinse, dry and polish
Process 3 ...				Final plate to required thickness	
Process 4 ...				Rinse, dry and polish	

71

TABLE 6

PLATING ON COPPER, BRASS, BRONZE AND NICKEL

Final Plate Required	Copper	Zinc	Pure Silver	Cupro-Nickel Hard Brass
Process 1 ...	No. 1 Copper, to required thickness	No. 1 Zinc, to required thickness	Quicksilver	Final plate direct, to required thickness
Process 2 ...	Rinse, dry and polish	Rinse, dry and polish	Final plate direct, to required thickness	Rinse, dry and polish
Process 3 ...			Rinse, dry and polish	
Process 4 ...				
Process 5 ...				

TABLE 7

PLATING ON LEAD, TIN, PEWTER, ZINC, BRITANNIA METAL AND SIMILAR ALLOYS OF LEAD

Final Plate Required	Copper	Zinc	Nickel	Pure Silver	Cupro-Nickel Hard Brass
Process 1 ...	No. 2 Copper, 20 minutes	No. 1 Zinc, to required thickness	No. 2 Copper, 20 minutes	No. 2 Copper, 20 minutes	Final plate direct, to required thickness, the current, being doubled during the first 3 minutes then reduced to normal
Process 2 ...	Rinse	Rinse, dry and polish	Rinse	Quicksilver	Rinse, dry and polish
Process 3 ...	No. 1 Copper, required thickness		No. 1 Nickel, to required thickness	Silver	
Process 4 ...	Rinse, dry and polish		Rinse, dry and polish	Rinse, dry and polish	

73

6 *CHROMIUM PLATING*

WARNING. The reader is referred to previous notes, in the latter part of Chapter 3, regarding the poisonous nature of chromic acid fumes. These fumes must not be inhaled, and when chromium plating is being carried on, either the fumes must be withdrawn by a suitable cowl and duct, or else the whole operation must be carried out in the open, or a well-ventilated shed. In the latter case, the operator must not remain in the shed longer than is absolutely necessary to make the connections and adjustments to the apparatus.

The deposition of chromium is considered by many experienced workers to be outside the scope of the amateur plater, but this view is based largely on the fact that huge currents are needed, and these are not readily available to the amateur. The author is of the opinion, however, that provided only small work is attempted, and that the reader is prepared to go to the trouble of being accurate, faithfully following the provisions outlined in this chapter, then successful chromium-plating is no more difficult than successful nickel-plating.

There is a point regarding supply of current for chromium plating, and this concerns the continuity of the current. The apparatus which has been described in Chapter 2, whilst giving a *direct* current, i.e., one which flows always in the same direction, does not produce a continuously flowing current. Those readers who are familiar with the properties of rectified A.C. current will realise that the output from the apparatus will consist of a pulsating current which alternately reaches maximum value, and sinks to zero, 100 times in each second (or, of course, 50 times if half-wave rectification is used).

To pursue this theory a little further, it can be said that the chromium electrolyte has a strong chemical action on the common metals of which the base might be composed, and this action is strongest when no current is flowing. Therefore, the opportunity for this chemical action occurs one hundred times per second, and the net result (without going any further into technical matters) is that the initial film of chromium is stripped off chemically almost as fast as it is laid in electrolytically. Consequently, the pulsating current is unsuitable for chromium plating, and some smoothing or " levelling " of the pulsations must be provided. This can be done in several ways, and the easiest and least expensive is that shown in Fig. 23, in which two large capacity condensers are connected in the rectifying circuit as shown. This will not produce an absolutely smooth current, except at very low current values, but if the capacity of these condensers is not below the value shown, they will prevent the current from falling to zero at any part of the cycle. An even simpler method, though not quite so effective, would be to connect one condenser of 1000mfd. across the output terminals of the panel. This would provide adequate smoothing up to a current of about 12 amps. on full-wave rectification. The condensers can be electrolytic, and of 25 volts rating, these being reasonably priced. The only point to watch carefully is that electrolytic condensers *must* be connected up the correct way as regards polarity (pos. and neg.).

The most satisfactory way of stabilising voltage is to use an accumulator across the output of the supply unit. A six-volt car battery is suitable, provided no more than six volts are used from the output. In cases where four or eight volts are required, it will be necessary to tap off a battery at the appropriate voltage, and before actually commencing plating under these conditions, it is strongly advisable

75

to check up, by means of an ammeter, that the battery itself is not taking any large current. Charge rate, as shown by the ammeter, should not exceed one or two amps. The current in the battery circuit should again be checked when the plating current is passing, to ensure that the battery is not discharging heavily. If it is, it should be tapped off at a lower voltage.

Accuracy is demanded at all stages of the process, including :

A. Original computation of surface area.

B. Making-up of the electrolyte.

C. Temperature of the electrolyte during plating.

D. Current-density.

E. Preparation, size, and disposition of the anodes.

SIZE OF JOB TO BE UNDERTAKEN

The apparatus described in the earlier chapters of this book, has been designed with a view to providing sufficient power to plate bases up to 160 sq. in. total surface area. This applies to the electrolytes so far described, but, owing to the peculiarly high currents demanded by the chrome bath, the size of base to be chromed, using the same apparatus, will be limited to about 16 sq. in.

FIG. 23a.

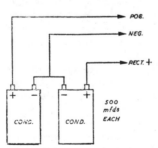

Arrangement of smoothing condensers for Chromium plating. Connections are : Pos. to output terminal positive, Neg. to output terminal negative, Rect. + to positive output lead from rectifier (R + on panel, see Fig. 6). The smoothing effect is least when least resistance is in circuit.

The reader will now have to consider whether he wishes to carry out chromium plating on a larger scale, and, if so, the size of all the apparatus will need to be adjusted accordingly, because, in order to deposit chromium on an area of 160 sq. in., a current of 120 amps., at 5 to 6 volts, will be required. The author, being concerned only with the chroming of small articles, is quite content with a 15 amp. output, and this allows about 20 sq. in. to be done.

The cost of the electrolyte being rather high, it is suggested that a very much smaller tank be used, especially in view of the small bases to be accommodated, and one which has given the author good service is a large enamelled iron saucepan, with the handle sawn off for convenience. The pan holds exactly one gallon, and is very convenient for small objects, a good deal of money thus being saved on the initial outlay for chromic acid.

If no gas supply is available, the pan and contents are placed in the empty plating tank, which is then filled with water almost to the height of the pan rim and the heater switched on. The connecting bars are moved, so as to be in the correct positions over the pan, and plating can then proceed. If the reader intends to heat the pan by gas, and use it separately from the plating tank, a connecting bar will have to be provided for the cathode, and can be of similar construction and fitting to those for the large tank. Anode bars are not required, as the anodes consist of strips of lead, and these can be bent, and hung over the sides of the pan, a small piece of felt, or cloth, being interposed between pan and anode to prevent any stray connections.

THE ELECTROLYTE

Pure Chromic Acid (dry crystals) ...	40 ozs.
Dilute Sulphuric Acid, B.P.	
(equiv. to 0.4 oz. wt. of pure acid) ...	4 ozs. (fluid)
Water to make	1 gallon

Plating Temperature: 45° Centigrade.

Current Density: 12 amps. per square decimetre.

This solution is corrosive and should not remain in contact with the skin. Cold running water will remove it.

In view of the very narrow margin of error allowed in making up this solution, the greatest care must be taken to ensure accuracy. There is no particular difficulty in dissolving the chromic acid, it being extremely soluble, and goes quite easily in cold water. The sulphuric acid is added last, and the whole well stirred. The correct proportion of sulphuric acid to chromic acid is very critical, and should be exactly 1 : 100. In such small quantities, the exact weighing of the sulphuric acid would be difficult, and, therefore, the recipe for the solution has been written to show 10% acid being used. This acid is obtainable from any pharmacy, accurately standardised to 10% content, and is known as Dilute Sulphuric Acid, B.P. Of this, ten times more will be required than of the pure concentrated acid, and it is conveniently measured by volume, instead of weight. If the reader does not possess a measuring glass, an ordinary medicine bottle may be used, each tablespoon marking being half a fluid ounce. These bottle-graduations are normally sufficiently accurate, but to make sure the amount measured should be checked in a second medicine bottle.

PREPARATION OF THE BASE

Copper, brass, nickel and iron can all be plated direct in the chromium bath, and of these, the three former usually are.

As, however, chromium gives protection only against tarnish, and not against rusting or corrosion, iron is, or should be, provided with a good undercoating of either copper or nickel before chroming, and of these, nickel is by far the most satisfactory. Copper, when laid directly upon iron, possess the peculiar property of accelerating rusting if any pin-holes are present in the copper plate, and if such a coppered iron base is chromed, the final plate has a nasty habit of peeling after a short time.

A further consideration with regard to iron bases, is the fact that chromium plate is so intensely hard that it is almost impossible to polish it, and hence all the polishing necessary must be carried out on the surface of the base, immediately before chroming. Iron is notably difficult to polish, and therefore the deposition of a plate of some other metal, which will take a high polish, between the iron and chromium, is indicated.

After a good deal of time spent in experimenting, the following procedures for cast and wrought iron, and for steel, have been evolved, and have proved very satisfactory.

1. The base is prepared as shown in Chapter 4, Table 3, as for nickel-plating, and is then put in the No. 1 nickel bath and plated for twenty minutes (cast and wrought), or fifteen minutes (steel) at the normal current.

2. It is removed, rinsed in water, and transferred to the No. 1 copper bath, plating again being carried on for forty-five minutes.

3. The base is then rinsed, dried and polished, and immersed in the following special dip for ten seconds :

Water	...	...	3 ozs.
Nitric Acid	...	...	1 oz.
Hydrochloric Acid	...	$\frac{1}{4}$ oz.

Mix together, and add slowly, with constant stirring :

Sulphuric Acid	...	...	5 ozs.

The dip is used cold.

4. The base is now rinsed very well, and returned to the nickel bath, where it is plated for a further thirty minutes.

5. Finally, a mirror surface is imparted to the nickel plate, and this is rather a tedious operation, as electro-deposited nickel is very hard. It is essential, though, that this polishing be very thorough, as any blemishes will be faithfully reproduced on the final chromium surface. All grease is removed from the surface, with hot soapy water or detergent and the actual chroming operation is carried out as follows.

DEPOSITION OF CHROMIUM

The special peroxidised lead anodes (to be described later) are arranged around the pan, according to the shape of work to be plated (see Fig. 23), making sure that they are all, as nearly as possible, equi-distant from the base, and also that the total area of anode immersed is not more than half the total base area.

The pan is filled with solution at the correct temperature, i.e., 45° Centigrade, the base connected to the cathode bar with a stout piece of flex, the current switched on, and then the base immersed quickly in the electrolyte.

The following tips regarding details of procedure will greatly simplify the operation of chroming.

1. When making up the anodes, a good length of flex should be allowed for each, and a connecting piece, as sketched in Fig. 24, is a decided asset in ensuring that all anode connections are good.

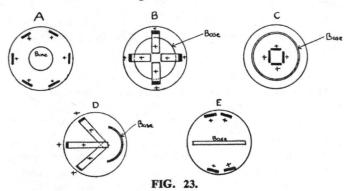

FIG. 23.

Disposition of anodes in Chromium plating.

A. *Cylinder, outside being plated.*

B. *Flat discs, upper side being plated. The anodes are bent inwards, horizontally, about 3 inches above the base. The under-side must be blocked out.*

C. *Hollow cylinder or similar vessel, inside being plated. The outside must be blocked out.*

D. *Reflector. Inside surface being plated; the rear surface must be blocked out. Anodes are bent as in* B.

E. *Rod or sheet. Whole surface being plated.*

It is merely a convenient bar of brass, drilled in six places along one side, to take the anode-wires, these each being held by small set-screws. At one end it is drilled for the main positive lead to the panel, and this also is held by a set-screw. Although a little time is required to make this, it is well repaid by convenience in use, and the certainty that all the anodes are well connected.

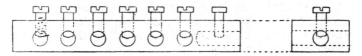

FIG. 24.

Connecting piece for anode-wires. The holes may be ⅛" and threads ³/₃₂".

Instead of such a connector, all the anode-wires could, of course, be twisted together and soldered, but this would involve always having all of them connected up at once, and, when plating small bases, some of them would be lying around, out of the solution. With the connector, only sufficient anodes need be connected for the particular job in hand.

2. Another good tip is concerned with the setting up of the anodes and the base in the pan. It is difficult to judge the length of cathode wire required, when the pan is filled with electrolyte, and the author invariably overcomes this by arranging everything whilst the pan is still empty. In view of the fact, mentioned before, that current must be flowing during the immersion of the cathode, the following routine will be found convenient and quite satisfactory :

81

The number and positions of anodes is worked out, and these are disposed around the empty pan, bent over the rim, and connected up. The base, with connecting flex attached, is connected to the cathode bar, the length of flex being adjusted so that the cathode hangs correctly. The solution, previously heated to the correct temperature, is then placed handy, the current switched on, and the solution poured quickly into the pan. If, as in the author's case, no other means of heating the solution are available, the cathode and anodes should be removed from the empty pan, in such a way that they can quickly be replaced in their respective positions, the pan filled with electrolyte, heated up with the tank heater, and then the anodes and cathode replaced, again making sure that the current is on before inserting the cathode.

After the correct length of time, which varies from five minutes to twenty minutes, according to the conditions to be withstood by the plating, the current is switched off, and the cathode immediately removed, followed by the anodes. They are all plunged into warm water, which is changed two or three times during ten minutes, to make sure than no chromic acid is left in contact, either with the anodes, or the plated article. The final rinsing water must be quite colourless.

The anodes are carefully dried, and put away until required again.

It will take a little practical experience before the panel controls can be set to pass the correct amount of current when the cathode is immersed. Trial pieces of brass or copper sheet, of the same area as the base to be plated, could be connected up, and the correct control setting noted. It will normally be found that about 6 volts is needed with very little resistance in circuit, and the ideal to aim at is a somewhat high "striking" current, for the first minute of immersion, which should then be reduced to the normal current, and plating carried on for the usual time. If the plant will supply sufficient current the striking could be done with 50% extra current, for forty-five seconds.

82

THE ANODES

The special peroxidised anodes, to which reference has already been made, are quite easily prepared. They consist of six strips of sheet lead, about 10 in. long and ½ in. in width. The thickness is immaterial, except in so far as strength is concerned. These are all scraped quite clean on both sides, using an old knife, and a length of 5 amp. flex is soldered to each. The best way is to tin the end of the flex for about ½ in., lay it across one end of the strip, solder in place, and then roll the end of the lead strip over, so that the soldered joint is protected by the bend. The fold can then be lightly hammered.

To peroxidise these anodes, they are all connected to the anode bar of the plating tank, and an equivalent sheet of lead is made cathode, the electrolyte being 5% sulphuric acid in water. A current of 5 amps. is passed for fifteen minutes, after which time connections are reversed, so that the lead strips are now cathode, and current again passed for fifteen minutes. Finally, the original connections are restored, and current passed for twenty minutes, the lead strips, at the end of this treatment, being covered with a hard, dark-brown coat of lead peroxide. This coating, whilst conducting well, resists the action of chromic acid, and prevents falling off in current during chromium-plating which occurs when clean lead is used, owing to the insulation of the anodes by a layer of lead chromate. They are finally rinsed in water and dried. If at any time during use, they should become spotted with yellow powder, it is a sign that the peroxide has become detached, and the peroxidising process should be repeated.

The commonest faults occuring during chromium-plating are three in number, and are as follows. (Temperature is assumed to be correct, i.e., 45° Centigrade.)

1. Milkiness in Deposit

This is a certain sign that current density is too low, and the remedy is obvious.

2. Dull, Grey Matt Deposit

The cause of this may be either too heavy a current, or else an excess of sulphuric acid. If the latter, it is somewhat difficult to correct, and can be satisfactorily done only by adding pure barium chromate, in successive small quantities, and trying a test-plate between each. If excess of barium chromate is added, the whole of the sulphuric acid will be removed, and this will be shown by a complete failure to deposit. When adding the barium chromate, not more than one-quarter of an ounce should be used, and this should be added at three or four stages, testing between each. The successful adjustment of sulphuric acid content, once it has become wrong, is a very difficult, and very tedious job. When the sulphuric content has been reduced to the correct level, the solution must be allowed to stand for a couple of hours, and then very carefully decanted off the sediment.

3. Complete Failure to Deposit

This may be caused by either a gross excess of sulphuric acid, or by insufficient, or else, in the presence of the correct amount of sulphuric acid, by impurities in the chromic acid. If, after carefully making up the electrolyte, and working it under correct conditions, no deposit is formed, or, at best, an indifferent one, then the purity of the chromic acid is definitely at fault; when purchasing this chemical, the highest quality must be demanded. That supplied by the retail chemist, if labelled " B.P." will be found quite satisfactory.

Before concluding this chapter, mention must be made of a peculiarity of the chromium electrolyte, that is, its property of improving with age. The first deposits obtained from a new bath are seldom excellent, although passably good, but an " aged " bath normally gives far superior plating, and the author makes it a rule always to age the electrolyte before undertaking any serious plating in it.

This ageing can be done in the cold, by inserting all the anodes, and a cathode of about 36 sq. in., say steel or wrought iron, and passing a current of approximately 10 amps. for about two hours per gallon of electrolyte. No agitation is necessary. At the end of this time, a dirty grey deposit of chromium will have been formed on the cathode, and if the reader wishes to be convinced of the difficulty of polishing chromium, he should try it on this deposit. None of the ordinary polishing compounds will touch it, and the commercial chrome polishes are very slow. Of the latter, that known as No. 5 white chrome is apparently the best for amateur use, and has some value for touching-up any patches of plate which are not quite perfect. The author, however, prefers to stick to the well-tried system of " polishing the chrome before it is put on."

Finally a word to those who, having followed the above instructions faithfully, have not succeeded in producing a satisfactory plate. This does happen sometimes, and there is invariably a definite reason for it. Maybe the chromic acid is not quite all that it should be, as regards purity. Maybe just a very little excess of acid has got into the solution, or maybe (and this is not impossible), the thermometer is not accurate. Temperatures throughout this book have been expressed in the Centigrade scale, and, of course, if a Fahrenheit thermometer is used, the necessary allowance must be made for the difference in the two scales.

Whatever the cause of an imperfect plate, it is always possible to make an improvement, and a little experimenting with temperature and current density will usually compensate for any slight errors in composition of the electrolyte, and vastly improve an indifferent chromium plate.

7 THE PLATING OF NON-CONDUCTORS AND ELECTRO-FORMING

THE deposition of metal on the surface of non-conductors is a very specialised branch of electro-plating, and has many commercial applications. From the author's point of view, however, it has been applied only to produce artistic effects upon glassware and china.

The methods which are to be suggested have been carried out on both the above bases, and are equally applicable to the numerous types of decorative plastics which are so popular at present.

As the principle of ordinary electro-plating depends upon the base having a surface which conducts the current to all its parts, it follows that any non-conductor, which is required to be plated, must, in the first place, be supplied with such a surface. As a matter of fact, the statement " electro-plating a non-conductor " is a misnomer, as it is not the non-conductor, but the artificial surface put on it, which is electro-plated. It also follows from this that the adhesion of the final plate depends entirely upon the adhesion of the conducting surface to the base. This point will be discussed under the various methods to be outlined.

Once the surface has been rendered conducting, it is a simple matter to build up this surface to any required thickness, with a deposit of whatever metal is required. There are three methods which the author has tried out, each one having been thoroughly tested, and a brief description of each, together with their advantages and disadvantages, will doubtless be of interest.

1. Graphite and Wax Method

This consists of painting the surface to be plated with a hot wax mixture of the following composition :

Paraffin Wax	1 part
Beeswax	2 parts
Powdered Graphite ...	2 parts

The waxes are melted together in a hot mortar, and the graphite thoroughly incorporated with a hot pestle. Whilst still liquid, the composition is cast into sticks, and allowed to cool. To use, the base is made hot enough to melt the compo, which is evenly applied, and the whole is then left to become quite cold. A little powdered graphite is then sprinkled on the surface, and brushed vigorously, using a very soft brush. The base is now ready for plating, and this can be done, either in the No. 1 nickel, or No. 1 copper baths; in either case, a very minute current is required at first, being gradually increased as the nickel or copper " flush " spreads over the work. A thick deposit of metal is laid on, followed by a rinse, and finally, whatever plate is required to finish.

The advantages of this method are that it is quick and cheap, and its main disadvantage lies in the fact that adhesion of the plate to the base depends solely upon the adhesion of the wax, and, therefore, no heat must be applied subsequently to the plating.

2. Mirror Method

In this, the original conducting surface consists of a mirror of silver, deposited on the base by chemical means, and this is then built up to any required thickness.

Numerous solutions have been devised, which will deposit silver on glass and similar surfaces, and the following is a very reliable formula. It is not an economical process, as silver is deposited not only on the surface required, but also upon the whole of the container in which the silvering is carried out. This, however, is unavoidable, and occurs with all similar methods.

87

The silvering solution is made up in the following parts :

A.	Rochelle Salts	...	...	2 ozs.
	Water to ...	...	...	8 ,,
B.	Strong Ammonia ...	...	1 oz.
	Water to ...	...	...	4 ozs.
C.	Silver Nitrate	...	...	$\frac{1}{4}$ oz.
	Water to ...	...	...	1 pint
D.	Caustic Soda	...	...	2 ozs.
	Water to ...	...	...	1 pint

To make the silvering solution, take the whole of solution C, and gradually add solution B, with constant shaking, yielding a brown precipitate. Continue adding solution B, constantly shaking, until the brown precipitate just barely dissolves. Next, add 1 oz. of solution D which again causes a brown precipitate, and again this is just dissolved by adding more solution B, but NOT in excess.

The solution as it now stands, will keep for a day or two, provided it is kept in a dark place. When it is required for silvering, the addition of solution A will cause all the silver to be deposited in about three hours. If only half of solution A is added, deposition will take about ten hours, but the deposit is finer.

The glass, or other base is, before dipping in the solution, prepared by thoroughly cleaning. This can be done by scrubbing with soap and very hot water, followed by a few moments in strong sodium cyanide solution, and after this, a good rinse in running water for five minutes, to remove the alkali. The actual surface to be plated must on no account be touched by hand after being cleaned. When ready, the base is immersed in the mixed solutions, preferably with the side required to be silvered uppermost, and left for three hours. It is then removed, rinsed rapidly, and transferred to the No. 1 silver bath, as cathode. Plating then proceeds to any desired thickness, interposing a layer of copper, if desirable for the sake of economy.

The advantage of this method is the excellent conducting surface provided by the mirror, rendering plating much easier, but the same disadvantage occurs, as with method No. 1, i.e., the adhesion is poor, and articles so plated will not stand up to much wear. The author has had some success in overcoming the adhesion problem, by coating the base with a very thin layer of gold size, and silvering just before it is quite hard. The mirror is then carefully, but thoroughly dried, and pressed firmly on to the backing of tacky size. After allowing a few days for hardening, the metal is built up in the plating bath, first with No. 1 copper, and following with quick-silver, and the final plate of silver. Some experience is required to guarantee success.

3. Gold Method

The final method to be described is the most expensive, but this is its only disadvantage, it being in all other ways exceptionally reliable.

It consists simply of painting the base with gold-size, allowing to become almost dry, and then gilding with leaf. This gives a very fine conducting surface, and any plate can be laid directly on the gold. It lends itself particularly well to the plating of intricate designs, such as monograms, etc.

Gold leaf is at present very difficult to obtain, and there are one or two substitutes which will answer quite well. Silver leaf is the best of these, and has only one disadvantage in that it tends to be somewhat brittle, and does not, therefore, give a continuous coat of metal over the surface, unless handled with great care when applied to the size. If silver leaf is employed, the first stages of plating must be very carefully watched, to make sure that all the surface is receiving a plate. This is shown by the surface becoming a uniform matt white all over, and, should any parts remain bright, it means that they are not receiving current. The remedy is to remove from the bath, dry carefully, and sprinkle a little " bronze " powder over the cracks in the surface, i.e., at the junction of the plated and non-plated portions. This bronze powder is brushed firmly into the cracks, with a fairly soft brush, and then quicksilvered as described previously. The base is then immersed, and plating recommenced. This will usually be found to correct the fault. If it fails, the whole should be scraped clean, and a fresh start made.

Plating on non-conductors, although not in any degree difficult, calls for the exercise of a little patience, and a great deal of care in manipulation. It is one of those crafts where an ounce of experience is worth a ton of instructions, and the novice should, therefore, not be put off by a few failures at first.

ELECTRO-FORMING

The process of electro-forming, that is the building-up of metallic parts or articles by means of the electric current is discussed in this chapter because one method is a modification of the plating of a non-conducting surface.

The amateur may find in electro-forming the answer to some of the awkward problems which occasionally crop up in the workshop when dealing with difficulties in sheet-metal working. By this process, intricate forms of sheet-metal work can be produced without the tedium of hammering, rolling, bending and fabricating which is often connected with that type of work. Its disadvantage lies mainly in the fact that it is somewhat slow, and may take upwards of a week if a fairly thick sheet of metal is required.

Briefly described, the process consists of fashioning a base or former of the required contours, and then plating a thick deposit of metal upon it, in such a way that the base can subsequently be removed, leaving only the deposited plate. The metal which lends itself best to this process is copper, as this metal can be deposited at a higher rate than any other. If more mechanical strength is required in the finished plate, there is no objection to making a composite or laminated deposit, such as alternating layers of copper and nickel, or brass may be introduced. It is recommended, however, that the initial plate be copper, as this is the best metal for giving a sound, even coating on the base, especially if the latter is non-conducting.

Non-Conducting Bases

These may be either wood or wax, or in some cases plastic. If wood or plastic, they must be of such a shape that they can be withdrawn from the sheath of metal which it is intended to form around them. For example, it is no use forming on a cylinder of wood, as it will be found impossible to withdraw the wood afterwards. If the "cylinder" is slightly conical, however, the wood will draw towards the larger end, and all such or similar types of former, in both wood or plastic, must have sufficient drawing angle. Where relief work is done on one surface only, the drawing angles must still be present, and there should be no vertical or under-cut surfaces.

These provisions do not, of course, apply to wax bases, as these can be removed by melting, and this constitutes the main advantage of a wax base. The disadvantage of all non-conducting bases lies in the difficulty in getting an original plate, and the only system worth trying is the wax process described previously. If the base is wood, it must be *soaked* in hot wax before applying the graphite-wax mixture, in order to waterproof it. Having done this, the graphite mixture is spread over the surface, and after allowing to cool, plenty of powdered graphite is brushed well into the surface, using a very soft brush. The base is then suspended in the No. 2 Copper bath, and plating commenced at a tiny current. When thoroughly covered with copper, current is increased to the normal value for this electrolyte, and continued for about ten or fifteen minutes.

At the end of this time, it is carefully inspected to ensure that it is completely covered. If any pinholes or bare patches are present, they should be well graphited, and plating continued. When coverage is complete, it is removed from the No. 2 Copper bath, *well* washed in cold running water, and placed in No. 1 Copper. Temperature

91

of this bath is raised to 35 deg. Centigrade, and three times normal current density is employed, i.e., about $4\frac{1}{2}$ amps. per sq. dm. Agitation must be very vigorous, and continuous throughout the whole plating time. Anodes should be of pure copper, and about the same area as the cathode.

Under these conditions, the copper plate will build up at the rate of about 0.0025 inches per hour. The plating is continued until the required thickness is attained, when the base is removed, washed well and dried. The former is then carefully removed, leaving the desired sheet of copper. Any other plate may be applied, either before or after removing the former, but it should be noted that in the case of nickel or brass forming, the normal current densities as given in Chapter 5 should not be greatly exceeded. This means of course that these metals give a much slower process than copper. For example, to build a 0.01 nickel plate would take approximately 40 to 50 hours.

Metallic Bases

When the use of low-melting point metal is possible in making the former, the process is rendered very much easier, as the excellent conducting surface of, say, tin-lead alloy makes the commencement of plating much simpler and much more reliable. A good alloy for making such formers is equal parts lead and tin, although plumber's solder, which contains much less tin, and is therefore cheaper, will answer quite well.

The former is either cast, machined, carved or built up in the low-melting point metal, and the surface is then made as good as possible, bearing in mind that imperfections in the surface will show in the finished plate. This base is then prepared as detailed in Chapter 4, Table 3 (Lead Alloys for Copper, etc.) and is immediately flushed over with copper in the No. 2 Copper bath. When thoroughly covered, it is washed and plating is continued in No. 1 Copper, as above.

When requisite thickness has been obtained, the job is again well washed, dried and heated until the metal runs out of the formed plate. In cases where the plate is continuous and in the form of a closed box or vessel, it will be necessary to make a hole at some part, to allow the molten metal to escape.

Points to Note

Electro-forming is, as mentioned previously, a somewhat slow process. It is not advisable to try and speed it up by using current densities higher than those detailed above. This might result in a spongy or badly formed plate, or in brittleness of the finished article.

Particular attention should be paid to the number and spacing of anodes, as bad arrangement of these will give a plate which is uneven in thickness and quality of metal.

A large tank should be used, so that outstanding parts of the base are not, relatively, much nearer to the anodes than recessed portions.

Electro-deposited copper is very soft, and this should be borne in mind when deciding upon the thickness required. The interposing of two layers of nickel each of 0.0005 inch, in a 15 thou. copper plate, has remarkable stiffening effect. and goes far towards producing a rigid article. However, there is a difficulty arising out of the effect of heat on bi-metallic articles, and it will be found that if the finished article is in the form of a bar or sheet, or if it has any large surfaces, there may be considerable bending or " dishing " under variations in temperature. It is inadvisable to construct a laminated article with large flat areas, especially if the final product is thin.

Finally, it should be remembered that unless electrolytic copper is used for the anodes in the No. 1 Copper bath, there will be some clouding of the solution after several hours' use, due to impurities in the anodes. When this occurs, plating should be held up for an hour or two, to allow the solution to settle. It may then be carefully decanted from the sludge, and operations continued.

93

8 *AN EXAMPLE*

A N earnest attempt has been made in the foregoing chapters to give all such information as will be required by the beginner, to enable him to become thoroughly acquainted with the subject, on the small scale. It often happens, however, in a work of this nature, that some small, but important detail of procedure is overlooked, the author's familiarity with the subject sometimes obscuring the fact that the beginner is not in a position to assume anything which ought to have been explained.

With the possibility of any such shortcomings in mind, it has been decided to devote this chapter to a detailed, stage-by-stage description of a reasonably easy exercise in plating. An attempt will be made to describe every detail of procedure, and where necessary, the exact reason for the various processes will be stated.

We will assume that the reader wishes to electro-plate the rim only of a model steam-engine flywheel, in cast iron, such as is sketched in Fig. 25. The dimensions given are, of course, arbitrary, and are merely for the sake of an example in the calculation of surface area.

The first job, if not already done, is to centre the wheel on a mandrel in the lathe, and put a good finish on the surfaces to be plated (shown shaded in the figure). After the final cut with the finishing tool, the lathe is run at full-speed, and the wheel polished as highly as possible, using grade " 0 " emery powder, mixed with machine oil, and applied with a leather pad. Sandpaper can be tried, but is not of much use for cast iron. Particular attention should be paid to the sharp edges, these being rounded just a little, as, otherwise, there is a tendency for the subsequent plate to be polished off during the finishing.

The next task is to calculate the surface area of the rim, the rest of the wheel being blocked-out, and this is done as follows.

Referring to the figure, it will be seen that the area of the outer edge of the rim is equivalent to a strip, whose area is :

$$\text{Circumference} \times \text{Width}$$

or $\pi \times 6 \times \frac{3}{4}$

$$= \frac{3.14 \times 6 \times 3}{4} \text{ which gives } 14.13 \text{ sq. in.}$$

The area of each face (on either side) is the difference between the areas of the larger and smaller circles, i.e., each face is :

$$(\pi \times 3 \times 3) - (\pi \times 2\tfrac{3}{4} \times 2\tfrac{3}{8})$$
$$= \pi (3^2 - 2\tfrac{3}{4}^2)$$
$$= 3.14 \times 1.44$$
$$= 4.5 \text{ sq. in.}$$

or   9 sq. in. for the two surfaces.

Adding this to the area of the outer edge, we have :

$$9 + 14.13 = 23.13 \text{ sq. in.}$$

As current densities are expressed throughout this book in terms of the square decimetre, the figure above is converted to these units by dividing by 15.7, i.e. :

$$\frac{23.13}{15.7} = 1.47 \text{ square decimetres.}$$

which for all practical purposes is $1\frac{1}{2}$ square decimetres.

The surface area having been calculated, the next step is to decide what the final plate shall be, and, in this instance, we will decide upon nickel. (Should chromium be required, it is quite easily laid on after the nickel.)

As the base is cast iron, the greatest care must be taken in preparing it for its plate, and a glance at Table 3,

Chapter 4, will show the sequence of cleaning operations to be carried out. Numbers 1 and 2 have been disposed of by polishing in the lathe, leaving the following yet to be done:

No. 4. De-greasing.

No. 5. Alkaline electrolytic cleaning.

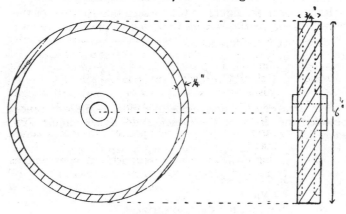

FIG. 25.

Cast iron flywheel; an example for plating.

At this stage, it is advisable to prepare all the solutions which are to be required, both for cleaning and for plating, as the job cannot be held up for any length of time whilst solutions are being made up. The alkaline cleaner is made up, and the author would make this up in a small container, just big enough to accommodate the wheel. For example, a small pan, about 7 in. diameter, and 5 in. deep would do excellently, and would hold about half a gallon. The quantities of chemicals needed to make this amount of No. 5 cleaner are only worth a few coppers, and the solution could be discarded after use.

It is assumed that the reader has five gallons of nickel No. 1 ready for the tank, and a supply of carbon tetra-chloride (almost all the non-inflammable cleaning fluids on sale in the shops consist of this). To proceed, then, the next item being the provision of connection to the base, and this should be done in such a way that the work need not be handled between cleaning and plating.

The following method is as good as any, and renders turning of the base during plating very simple. A length of screwed brass rod is inserted through the centre hole of the wheel, nuts and washers fitted at each side, the rod accurately centred, and the nuts screwed up tight. The screwed rod should project $\frac{1}{2}$ in. one side, and 1 in. on the other, this longer portion serving as a handle, when the wheel has to be held in the hand. It should be nicely balanced on the centre rod, and come to rest in any position. The reason for this will be apparent when the actual plating is described.

The cleaning processes are now carried out, commencing with the degreasing. This is to remove all traces of the oil used in polishing in the lathe, and, as stated in Chapter 4, consists in rubbing the whole of the wheel with pieces of clean cloth, liberally soaked in tetrachloride. The piece of cloth is renewed three times, with fresh application of the cleaner, and allowed to dry. Next, some very hot water is procured, and the job immersed in it, to warm it. Using household detergent, and a clean nail-brush, the whole of the surface is now scrubbed up to a good lather and rubbed vigorously for a few minutes. Soap should not be used.

Having well scrubbed the job, it is rinsed in hot water. The longer end of the centre-rod is now fitted with two washers, holding a length of bared 20 s.w.g. copper wire, and a further nut tightened up. From this point onwards, the hands must not be allowed to touch those parts to be plated, and so, holding it by the wire, it is lowered into the alkaline cleaner, and left there whilst a sheet of steel about 4 in. × 4 in. is fitted with a connecting wire, and securely connected to the Pos. output terminal on the panel. A temporary frame is fitted over the pan, to allow suspension of the steel anode in the liquid, about half an inch below the surface, and in a horizontal plane, i.e., facing the wheel, which is lying more or less flat on the bottom. The steel anode should be a little to one side of centre, so that the wire from the base can pass out without shorting. The pan, containing base, is now heated to 75° Centigrade, and all is ready for the electrolytic cleaning.

The transformer tapping is set at 6 volts, and the whole of the resistance put in circuit. The job is connected to the Neg. output terminal, and, after making sure that no shorting can occur in the pan, current is switched on. If both switches over the ammeter are open, a reading of about $\frac{1}{4}$ amp. will be obtained, and the resistance is then reduced until a current of 3 amps. is obtained. This corresponds with 2 amps. per square decimetre, as laid down in the description of the process. This current is passed for five minutes, at the end of which time the wheel is carefully taken out of the bath, and inspected. It should appear quite bright, and the whole of the surface should now be wet. If any waxy-looking patches remain, a few more minutes are allowed at a somewhat higher current, say 2 amps. with the " X2 " switch closed, which corresponds with 4 amps. actual current. When the operator is satisfied that the surface is clean, current is switched off, and the anode removed, the base being left in the solution whilst the proper plating tank is got ready.

This latter is filled with the No. 1 nickel, and two anode bars fitted (one at each end of the tank). These are connected together, and to the Pos. output on the panel, and the cathode bar is connected to the Neg. The base is carefully lifted out of the pan, and placed in running water for five minutes. It is of the first importance that all alkali should be rinsed off the base or else it will upset the acid balance of the nickel electrolyte.

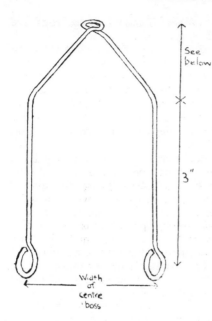

FIG. 26.

Sling for flywheel. The distance from connecting loop at the top, to the bend in the wire, is determined by the depth of the tank, and height of the tank-bars above the liquid.

The blocking out had perhaps better be tackled next, and this can be done in various ways. With such a base, the author would hold it horizontally in the vice, by gripping one end of the screwed rod in the jaws, so that the upper face is quite level. The cellulose lacquer is then applied quickly and evenly to the upper surface and a few minutes allowed for it to set. Care should be taken not to apply it so freely that it runs on to any surface which is to be plated. The whole is then turned over and the reverse side treated in the same way. Then leave for about an hour in a warm room, to allow the lacquer to harden.

Meanwhile the nickel anodes can be placed in position and connected up to their respective bars. Various shapes and sizes of nickel rods are available, and the author uses

$\frac{1}{2}$ in. diameter round. When obtained, these rods are usually very dirty, and a good cleaning with a smooth file is necessary. The reason for this is that when the bath is working properly the nickel should dissolve off the anode at the same rate as it is being deposited on the cathode, and this cannot occur if the anodes are not properly clean.

For the job in hand, two $\frac{1}{2}$ in. bars would be sufficient, as it is technically advantageous, when using the No. 1 nickel bath, to have the anode area somewhat less than that of the cathode.

If the anodes are of small diameter nickel rod, say $\frac{1}{4}$ in., it would be advisable to use four, one at each corner of the tank, and in adjusting the position of each, the agitator shaft and propeller must be considered, and a reasonable amount of space allowed, to avoid fouling of the latter by the anodes. The anodes themselves, whatever their diameters, must be long enough to come out of the solution at the top, so that the copper connecting wires shall not be under the surface. These latter can be joined to the anodes in any way, provided that good connection exists; the author always uses soft-solder.

A glance at Chapter 5 will show that the plating current required for this electrolyte is $\frac{1}{2}$ amp. per square decimetre, and this means that a current of $\frac{3}{4}$ amp. is correct for the job in hand. After some experience, it is easier to judge how the controls should be set, for various electrolytes, and various sizes of base. to obtain nearly the correct current straight away, but until such experience is gained, the beginner is advised always to use a low voltage, either 4 or 6 volts, and to set the resistance at maximum before switching on the current. In this particular case, 4 volts will usually do, provided that the rectifier has not a large internal resistance, and the transformer switch is, therefore, set at the 4 volt tapping, with the resistance over at the maximum side.

The base now requires a copper sling, of the shape shown in Fig. 26. This is carefully adjusted so that it will hold the wheel in a true vertical plane, and again, whilst

100

adjusting, the wheel itself must not be handled, only the screwed rod being touched. When the sling is ready, it is fitted to the connecting screw on the cathode bar boss and the wheel slipped in, after which the cathode bar is placed in position on the tank. All is now ready, and current is switched on, when a very small reading will be obtained on the ammeter. This is brought up to $\frac{3}{4}$ amp., by adjusting the resistance switch, and, if it cannot be raised to this value, the switch arm is put back to the maximum position, the voltage increased to six, and current again adjusted.

When the correct current is obtained, the agitator motor is connected up, and plating can proceed. A further reference to Chapter 5 will show that it takes half an hour to deposit one ten-thousandth of an inch of plate, and from this the total time to deposit the required thickness can be calculated. For such purposes as the present base is required, half a thou. will be amply sufficient, this taking two and a half hours, and little need be done during this time, except occasionally to revolve the wheel through about one-quarter of its circumference. The reason for this occasional disturbance of its position is the fact that the parts of the surface immediately behind the sling-wires are " shadowed " by them, and the wires receive nickel plate at the expense of the wheel. This occasional movement ensures that all parts of the rim have an equal chance of receiving the deposit. In the position in which the wheel is suspended in the tank, it will be seen that the flat surfaces, on each side, are more likely to receive deposit than the outer surface, and, in practice, the plate on these will be about 10% thicker than on the curved surface. If this is considered to be a disadvantage, the answer is to sling the wheel in a plane running the length of the tank, and, in this case, it must be revolved about its axis very frequently, i.e., a quarter turn every fifteen minutes.

At the end of the time allowed for plating, current is switched off and the work removed from the tank. It is disconnected, and put to soak in cold water, whilst the anodes are removed, washed and put away. When thoroughly rinsed, the wheel is dried and the cellulose removed by either dipping in thinners and allowing to soak, or by repeated rubbing with a rag soaked in thinners, until the lacquer is removed. This may take some time, and it may be considered that as the unplated portion is iron, and likely to rust, the lacquer could perhaps be left on as a protection against this.

All that remains now is the final polishing, and an examination of the surface of the nickel may reveal the presence of a few rough patches, usually of a darkish grey colour. These, whilst capable of being polished out on the buff, are much more easily treated by a very light preliminary rub with the finest grade sandpaper, until a fairly bright surface is obtained. This is then followed by a good buffing, running the calico wheel at high speed, as a fair amount of pressure is needed. Any of the usual grades of polishing compound are suitable, but the author always favours the red variety, especially for nickel.

Finally, the centre of the wheel is enamelled, or left bare, according to the requirements of the model. If it is to be left bare, it may be found that the nickel has over-run its proper boundary, owing to the fact that the blocking-out was not carried quite up to the line. In this case, a very light cut, with a very sharp tool, in the lathe, will restore the sharp dividing-line.

Just one final word. If the reader should decide on putting a chrome plate over the nickel, it must be done as soon as polishing is finished. Even by the next day, the nickel will have acquired a thin covering of oxide, which, though quite invisible, will yet be just sufficient to spoil the perfect adhesion necessary between the two adjacent metals, and, of course, if chrome is to be added, the removal of blocking-out will be delayed until the final plate is finished.

9 ANODISING

A LTHOUGH the process of anodic oxidation of aluminium, and its alloys, cannot actually be described as electro-plating, it has been thought fit to include a chapter upon this subject, as it is carried out with the aid of the same apparatus as is electro-plating proper.

Briefly, the process consists of depositing a film of aluminium hydroxide on the surface of the base, in order to protect it from irregular corrosion. The protection conferred by anodising is quite good, and will resist atmospheric corrosion under normal indoor conditions. It is much less permanent out-of-doors, being quickly destroyed by rain or frosts.

Its main value, however, lies in the peculiar property of the anodic film to absorb dyes, by which means many highly decorative effects can be obtained. The dyed surface is quite "fast," and will withstand boiling water, without loss of colour. The process is applicable to pure aluminium, and to alloys with a high aluminium content, providing no copper is present in the alloy, and is carried out as follows.

The article is highly polished, using a fine grade compo, and is then de-greased with carbon tetrachloride. This is followed by a washing in hot detergent, and rinsing well in clean cold water. Being thus prepared, the base is hung from the anode bar of the tank, using pure aluminium connectors, and two or more lead plates are hung from the cathode bars, this time using lead connectors, or of course, lugs projecting from the lead sheets. It is advantageous to employ cathodes on both sides of the job, to ensure an even coating of oxide, and the total cathode area should be about twice the base area, though this is by no means critical.

The tank is now filled with the following electrolyte :

Sulphuric Acid (by volume)   5 ozs./gal.

Glauber Salts (Sodium Sul-
phate)   2   ,,

Method : Dissolve the sodium sulphate in three-quarters of the water, add the sulphuric acid slowly, and with constant stirring. This will generate heat, and the solution should be cooled, and made up to volume with water.

This electrolyte is used cold, i.e., from 12 to 20° Centigrade, and the current should be in the neighbourhood of 2 amps. per square decimetre. About 10 volts will be required, and the current will have to be adjusted after about twenty minutes, as it will gradually decrease with increase of oxide on the anode. Total time required is about thirty to forty minutes, which gives a good solid facing of oxide, and, at the end of this time, the work is withdrawn from the tank, and rinsed in cold water. The appearance of the surface at this stage is not very much different from the original, being, perhaps, a little cloudy, and the final step in the process is to " fix " the film.

This is carried out by boiling the base in clean water for ten minutes, which converts the oxides into very complex hydroxides, these latter forming a vitreous, protective coating, and this being transparent, allows the polished surface of the metal to shine through. A rinse in hot water, and a thorough drying, complete the process. If a dyed surface is required, however, the dye should be added to the water in which the base is boiled, as described above, and in this case a further boiling in clean water is advisable, in order to remove any dye which has not been " fixed " in the anodic film.

The above details outline the necessary steps in the anodising process, in its simplest form. There is, however, a modification of it which might now be discussed. This concerns the use of alternating current, which, whilst not possessing any advantage with regard to quality of film produced, has, nevertheless, decided advantages in the simplicity of equipment required. The process can, indeed, be quite efficiently carried out with the aid of a transformer and ammeter only. Such a transformer will require secondary tappings of 16, 20 and 25 volts, and be capable of delivering a sustained alternating current of say 6 amps.

The process is as follows. Instead of using a lead cathode, either an aluminium one is used, or else the articles to be treated are divided into two equal groups, and these groups used as opposing electrodes. (The terms " anode " and " cathode " do not, of course, apply in the case of A.C.) The groups of bases are hung on two tank bars and connected directly to a suitable tapping on the transformer, i.e., one which gives approximately the correct current, using, of course, an A.C. ammeter.

Composition of the special electrolyte is :

Pure Oxalic Acid ...   ...   10 ozs./gal.

Method : Dissolve in half the amount of boiling water, in a non-metallic vessel, and allow to cool. Make up to volume with boiled and cooled water. With water from limestone districts, or containing magnesium salts, a deposit will form, and this should be allowed to settle, and the clear liquid poured off for use.

Operating temperature is 35° Centigrade (blood-heat) and current is best kept around 1 amp. per square decimetre. Here again, the current tends to fall off, and after about fifteen minutes it will be necessary to increase the voltage. As oxidation of each group of bases only takes place on half of each cycle of current, a much longer time is required, and about ninety minutes will be necessary to give a satisfactory surface.

Subsequent rinsing and boiling is required, and is carried out exactly as described for the sulphuric process.

105

The coating produced by this A.C. method is not quite so transparent as the sulphuric one, but will take dyes just as readily, giving a rather deeper colour with similar dye.

Finally, it must be explained that no metal, other than lead and aluminium, should be allowed in contact with either of the anodising electrolytes, at any time, and in all cases, the base must be suspended from the tank bars by strips of pure aluminium only.

The correct dyes to use are best found by trial and error, there being so many dye-stuffs on the market at present. The author has had a good deal of satisfaction in using many of the well-known brands of household dyes, and recommends a little experimenting in order to ascertain the correct strength at which the dye solutions should be used. Broadly speaking, good results are obtained by mixing the dye as if intended for cotton material, but no other substance (e.g. vinegar etc.) should be added to the dye, even if recommended by the makers. The reason for this lies in the fact that aluminium hydroxide is its own mordant, or dye-fixative, and the addition of acids or salts to the dye bath might easily impair the anodic film.

The wearing properties of the film, whether dyed or not, are much inferior to a metallic surface, and owing to the thinness of the film, scratches cannot be polished out. This explains the reason for polishing the base prior to anodising, and also gives rise to the custom of lacquering the final product, if it be required to stand up to light wear and tear. Of the numerous lacquers available, those of the cellulose group are by far the most suitable. The film itself provides an excellent surface for holding a cellulose lacquer, which is best applied either by spraying or by dipping. There are many reliable brands of clear, colourless cellulose available, and all appear to give satisfaction, if carefully applied.

Of all methods for protecting and improving the appearance of aluminium, the author considers that anodising is by far the simplest, and by no means the least pleasing, when carefully carried out.

106